Forgotten Algebra:
A Self-Teaching Refresher Course

Second Edition

BARBARA LEE BLEAU, Ph.D.

Jefferson-Pilot Professor
of Management Science
Division Head
Business and Economics
Pfeiffer College
Misenheimer, North Carolina

Barron's Educational Series, Inc.

All inquiries should be addressed to:
Barron's Educational Series, Inc.
250 Wireless Boulevard
Hauppauge, New York 11788

International Standard Book No. 0-8120-1942-3

PRINTED IN THE UNITED STATES OF AMERICA
23 22 21 20 19 18 17 16 15 14 13 12

Preface

Forgotten Algebra is for the person with a deficiency in algebraic skills, whether this deficiency has resulted from inadequate learning or from the passing of time. The workbook was prepared originally to provide a means of removing such deficiencies among college students (especially returning adults) without hindering their progress by requiring them to take remedial courses.

Forgotten Algebra was designed as a teach-yourself text-workbook. It can supplement various traditional mathematics courses, or it can be used to "brush up" before entering a course or taking a standardized entrance examination such as the SAT, GRE, or GMAT. Alternatively, if you are an adult who has never studied algebra but now feel the need for knowledge of this subject, here is an excellent introduction.

Each unit of the workbook provides explanations and includes numerous examples, problems, and exercises with detailed solutions to facilitate self-study. Students should be able to complete all units satisfactorily in one semester in conjunction with a full load of courses.

Readers have found *Forgotten Algebra* helpful with a variety of quantitative courses ranging from economics to engineering, as well as for review before entering graduate studies. The workbook has been described as easy to read, yet containing "all the algebraic information a person needs but is apt to forget."

June 1994

Barbara Lee Bleau
Misenheimer, North Carolina

CONTENTS

OTHER BOOKS CITED IN TEXT

Drooyan, Irving, and William Wooton, *Programmed Beginning Algebra*, 2nd Edition, Vols. I–V, John Wiley & Sons.

Leithold, Louis, *College Algebra*, Macmillan Publishing Company.

Lial, Margaret L., and Charles D. Miller, *College Algebra*, Scott, Foresman and Company.

Peters, Max, *College Algebra*, Barron's Educational Series, Inc.

Rees, Paul K., Fred W. Sparks, and Charles Sparks Rees, *Algebra, Trigonometry, and Analytic Geometry*, 2nd Edition, McGraw-Hill Book Company.

Rich, Barnett, *Elementary Algebra*, Schaum's Outline Series in Mathematics, McGraw-Hill Book Company.

UNIT 1

Signed Numbers

In this first unit of your workbook you will learn the meanings of the terms *signed number* and *absolute value,* and the rules for performing the four basic operations of addition, subtraction, multiplication, and division using signed numbers.

TWO IMPORTANT DEFINITIONS

We will start with two important definitions.

> Definition: A **signed number** is a number preceded by a plus or minus sign.

A number preceded by a minus sign is said to be a **negative number**. For example, –25 is a negative number.
A plus sign is used to denote a **positive number**; +5 is a positive number.
If no sign is written, the number is understood to be positive; 20 is a positive number.
Zero is an exception. The number 0 is neither positive nor negative.

> Definition: The **absolute value** of a signed number is the number that remains when the sign is removed.

For example, the absolute value of –13 is 13. When the negative sign is removed, the number that remains is 13. Likewise, the absolute value of +2 is 2.
Instead of writing the words *absolute value*, we can use the symbol | |. Thus |–7| is read as "the absolute value of negative 7." And |–7| = 7.

Before continuing, here are a few problems for you to try.

Problem 1   Find the absolute value of –37.

Problem 2   Find $|+5|$.

Problem 3   Find $|-11|$.

Problem 4   Find $|0|$.

Answers: 1. 37   2. 5   3. 11   4. 0

ADDING SIGNED NUMBERS

Rule 1:   **To add numbers with like signs, add their absolute values. To this result prefix the common sign.**

EXAMPLE 1 $(-5) + (-3)$

Solution: Add the absolute values 5 and 3. To the result, 8, prefix the common sign. Thus, $(-5) + (-3) = -8$.

EXAMPLE 2 $(+5) + (+3)$

Solution: Use the same reasoning as for Example 1, but the common sign now is plus. Thus, $(+5) + (+3) = +8$.

Rule 2:   **To add two numbers with unlike signs, subtract the smaller absolute value from the other. To this result, prefix the sign of the number having the larger absolute value.**

EXAMPLE 3 $(-5) + (+3)$

Solution: Subtract the absolute value 3 from the absolute value 5. To the result, 2, prefix the sign of the number having the larger absolute value. Thus, $(-5) + (+3) = -2$.

Now you try a few.

Problem 5 $(+5) + (+7)$

Problem 6 $(-9) + (-2)$

Problem 7 $(-1) + (-2) + (-3)$

Problem 8 $(-10) + (+2)$

Problem 9 $(7) + (-3)$

Problem 10 $(-2) + (+8)$

Problem 11 $-1 + 7$

Problem 12 $12 + (-1)$

Answers: 5. 12   6. –11   7. –6   8. –8
9. 4   10. 6   11. 6   12. 11

I hope you noticed that, when adding two signed numbers, you add absolute values for like signs and subtract absolute values for unlike signs.

I fully realize you could have gotten all of Problems 5–12 correct, without ever knowing the two rules, by simply using a calculator. And if the numbers had been more complicated, say $(-1237.15) + (352.79)$, I expect that's what I would have done—used my calculator. But for easy numbers, like those above, it's faster to do them in your head than to rely on a calculator.

SUBTRACTING SIGNED NUMBERS

Rule 3:   **To subtract a signed number, add its opposite.**

EXAMPLE 4 $(5) - (-2)$

Solution: To subtract –2 from 5, the rule says to add the opposite of –2. The opposite of –2 is 2.
Thus, $(5) - (-2) = 5 + 2 = 7$.

EXAMPLE 5 $(-6) - (-2)$

Solution: The opposite of –2 is 2.
            Thus, $(-6) - (-2) = -6 + 2 = -4$.

EXAMPLE 6 $(-3) - 6$

Solution: The opposite of 6 is –6.
            Thus, $(-3) - 6 = (-3) + (-6) = -9$.

In each of the above examples, the subtraction was changed to addition by rewriting the problem with the signed number's opposite. Then the final answer was found by using the rules for the addition of signed numbers.

Here are some problems for you. Try them without using a calculator.

Problem 13 $23 - 18$

Problem 14 $5 - (17)$

Problem 15 $(-3) - (2)$

Problem 16 $2 - (-7)$

Problem 17 $-10 - (-5)$

Problem 18 $-11 - 2$

Answers: 13. 5   14. –12   15. –5   16. 9   17. –5   18. –13

MULTIPLYING SIGNED NUMBERS

Rule 4: **To multiply two signed numbers with like signs, multiply their absolute values and make the product positive.**

Rule 5: **To multiply two signed numbers with unlike signs, multiply their absolute values and make the product negative.**

EXAMPLE 7 $(5)(2) = 10$
EXAMPLE 8 $(-3)(-5) = +15$
EXAMPLE 9 $(-1)(+7) = -7$
EXAMPLE 10 $(+3)(-2) = -6$

Think of it like this: Two like signs yield plus, and two unlike signs yield minus. Now try the following problems.

Problem 19 (–3)(5)

Problem 20 (–2)(–2)

Problem 21 (–4)(–11)

Problem 22 6(–8)

Problem 23 (–1)(7)

Problem 24 (4)(–3)

Answers: 19. –15 20. 4 21. 44 22. –48 23. –7 24. –12

DIVIDING SIGNED NUMBERS

Rule 6: **To divide two signed numbers with like signs, divide their absolute values and make the quotient positive.**

Rule 7: **To divide two signed numbers with unlike signs, divide their absolute values and make the quotient negative.**

EXAMPLE 11 $-35/5 = -7$

EXAMPLE 12 $-63/-7 = 9$

EXAMPLE 13 $10/-5 = -2$

EXAMPLE 14 $-12/-2 = 6$

You should now have an understanding of signed numbers and absolute value and be able to add, subtract, multiply, and divide signed numbers.

Briefly stated, the rules are as follows:

Addition: When the signs are the same, add the absolute values and keep the common sign. With unlike signs, subtract the absolute values and keep the sign of the number with the larger absolute value. (Rules 1 and 2)

Subtraction: Change subtraction to addition of the opposite, and follow the rules for addition. (Rule 3)

Multiplication and division: Two like signs yield a plus; two unlike signs yield a minus. (Rules 4–7)

Before beginning the next unit, try the following exercises without using a calculator. When you have finished, check your answers against those at the back of the book.

EXERCISES

Perform each of the indicated operations.

1. $-2 + (-5)$
2. $1 - (-3)$
3. $8 - 17$
4. $-4 - (-3)$
5. $6(-8)$
6. $(-3)(6)$
7. $-4(-11)$
8. $32/-8$
9. $-25/-5$
10. $-3 + |-4|$
11. $|-6| - 7$
12. $90/-30$
13. $-11 + 20$
14. $-6 - (-3)$
15. $-3 + 2 + (-5)$
16. $-5 - 5$
17. $(-2)(-3)(-4)$
18. $(-4)(-8)$
19. $-3 + 1 - (-2)$
20. $(-1)(-1)(-1)(-1)$

UNIT 2

Some Important Definitions, Grouping Symbols and Their Removal, and Simplifying Expressions

In this unit you will learn the meanings of some of the words we will be using throughout the course. Then you will learn about various grouping symbols and the way they are removed to simplify equations.

SOME IMPORTANT DEFINITIONS

Five of the most important concepts you will need in the course are:

1. Term
2. Variable
3. Coefficient
4. Expression
5. Like terms

Here are some "commonsense" or "working" definitions of these concepts. It is *not* important that you memorize these definitions, but you should understand each one and be able to define the concept in your own words.

> Definition: A **term** is a symbol or group of symbols separated from other symbols by a plus or minus sign.

For example, in $3x - 5y + 7xyz$, there are three terms. Note that terms may contain several numbers and letters. Therefore $7xyz$ is a single term; the number is written first, and the letters usually are put in alphabetical order.

> Definition: A **variable** is a letter or symbol used to represent some unknown quantity.

For example, x is often used as a variable. In the term $2y$ the variable is y. Variables with exponents such as x^2 and y^3, will be introduced in Unit 8.

> Definition: The **coefficient** of a variable is everything else in the term except the variable.

In the term $2y$ the coefficient of y is 2. Here are some other examples:
a. x The coefficient of x is understood to be 1.
b. $7ax$ The coefficient of x is $7a$.
c. $5z$ What is the coefficient of z?

> Definition: One or more terms may be referred to as an **expression.**

Expressions are given different names depending on the number of terms involved. For example, expressions with one term are called monomials, while expressions with two terms are called binomials. The expression $3x - 5y + 4x$ has three terms and is called a trinomial.

Definition: **Like terms** contain the same variable and differ only in their coefficients.

In the expression $3x + 4x$ the terms are like terms that differ only with respect to their coefficients, 3 and 4. Like terms can be combined. The terms of the expression $3x + 4x$ can be combined to form a single term, $7x$. Similarly, like terms in the trinomial expression $3x - 5y + 4x$ can be combined, forming a binomial expression, $7x - 5y$.

Try a few problems to test your understanding of these important concepts.

Problem 1   How many terms are there? $10zy + 3x - 5an + 12$

Problem 2   Which is the *variable*? $7a$

Problem 3   What is the *coefficient* of y? $5aby$

Problem 4   A binomial expression has _____ terms.

Problem 5   When the like terms in the expression $10y + 8z + 3y$ are combined, the result will be _____ .

Answers: 1. four   2. a 5. $5ab$ 4. two   5. $13y + 8z$

GROUPING SYMBOLS

Now let's turn our attention to grouping symbols. There are three common types of grouping symbols:

parentheses	( )
brackets	[ ]
braces	{ }

Whether you use parentheses, brackets, or braces, these symbols indicate that the quantities enclosed within them are considered to be a single unit with respect to anything outside the grouping symbol. In the expression $8y + (3a - b) + 2x$, the terms $3a - b$ are considered as a single unit, separate from the terms outside the parentheses.

REMOVING GROUPING SYMBOLS AND SIMPLIFYING

To remove grouping symbols we use the **distributive property**:

$$a(b + c) = ab + ac$$

EXAMPLE 1 $2(x + y) = 2x + 2y$

EXAMPLE 2 $-3(x - 5) = -3x + 15$

Note the plus sign between the terms, since $(-)(-) = +$.

Now consider some more complicated expressions. We will **simplify** these expressions by removing grouping symbols and then combining like terms.

EXAMPLE 3

Simplify: $2(4xy + 3z) + 3[x - 2xy] - 4\{z - 2xy\}$.

Solution: $2(4xy + 3z) + 3[x - 2xy] - 4\{z - 2xy\}$

$$= 8xy + 6z + 3x - 6xy - 4z + 8xy$$

Note plus sign.

$$= 3x + 2z + 10xy$$

EXAMPLE 4

Simplify: $3aw + (aw + 4z) - (2x - 3y + 4z)$.

Solution: $3aw + (aw + 4z) - (2x - 3y + 4z)$

$$= 3aw + aw + 4z - 2x + 3y - 4z$$

$$= 4aw - 2x + 3y$$

Actually we are multiplying by 1, which is understood to be there but is not written.

Try the next one on your own.

Problem 6

Simplify: $2(x - 1) - 3(2 - 3x) - (x + 1)$.

Solution:

Answer: $10x - 9$

- -

Now for more complicated expressions!

> When one set of symbols is within the other, the grouping symbols must be removed from the inside out.

EXAMPLE 5

Simplify: $x - [5 - 2(x - 1)]$.

Solution: $x - [5 - 2(x - 1)]$

$$= x - [5 - 2x + 2]$$
$$= x - [5 - 2x + 2]$$

$$= x - 5 + 2x - 2$$
$$= 3x - 7$$

Now you do one.

Problem 7

Simplify: $-3(a + b) - 2(a - b) + 5[a - 2(b - a)]$.

Solution:

Answer: $10a - 11b$

Did you get the right answer? Let's do one more before we quit.

EXAMPLE 6

Simplify: $3\{x - 2(5 - [x + y]) + 1\} - 2$.

Solution: $3\{x - 2(5 - [x + y]) + 1\} - 2$

$$= 3\{x - 2(5 - x - y) + 1\} - 2$$

$$= 3\{x - 10 + 2x + 2y + 1\} - 2$$

$$= 3x - 30 + 6x + 6y + 3 - 2$$

$$= 9x + 6y - 29$$

You should now be able to define, in your own words, the basic concepts *term, variable, coefficient, expression,* and *like terms*. These basic concepts will be used continually throughout the following units.

You should also be familiar with three common types of grouping symbols. Remember that the quantities enclosed within these symbols are considered to be a single unit, separate from anything outside the parentheses, brackets, or braces.

Remember also that, when one set of symbols is within the other, grouping symbols are removed by working from the inside out. Simplifying then involves combining the like terms.

Before beginning the next unit you should simplify the following expressions. When you have completed them, check your answers against those at the back of the book.

EXERCISES

Simplify:

1. $(7x - 2y + 5) + (2x + 5y - 4)$
2. $(3x + 5xy + 2y) + (4 - 3xy - 2x)$
3. $(5y - 2a + 1) - (y - 3a - 7)$
4. $3(2a - b) - 4(b - 2a)$
5. $2(7x - 5 + y) - (y + 7)$
6. $5 - 2(x + 2[3 + x])$
7. $x - \{2x - 2(1 - x)\}$
8. $4x - \{3x - 2[y - 3(x - y)] + 4\}$
9. $2x - [2x - (-x - \{1 - x\})]$
10. $-3a - 5\{2a - 2[2a - (4a - 2) - 2(7 - 3a)]\}$

If additional practice is needed:
Drooyan and Wooton, Unit 4, page 60, exercise 1
Leithold, page 59, problems 13–22
Peters, page 23, problems 1–10
Rees, Sparks, and Rees, page 50, problems 25, 26, 28–30, 35, 36
Rich, page 68, problems 19, 27

UNIT 3

Solving First-Degree Equations

The purpose of this unit is to provide you with an understanding of first-degree equations. When you have finished the unit, you will be able to identify first-degree equations, distinguish them from other types of equations, and solve them.

What is a first-degree equation? A **first-degree equation** has these characteristics:

1. There is **only one variable**.

2. The variable is involved in **one or more of only the four fundamental operations** of addition, subtraction, multiplication, and division.

3. The variable is **never multiplied by itself**.

4. The variable does **not** appear **in any denominator**.

Here are some examples of first-degree equations:

$$2x + 3 = 15$$

$$\sqrt{5}x - \frac{x + 2}{2} = \pi$$

$$3(x - 1) + 2[5 - 3(x + 4)] = 0$$

Here are some examples that are *not* first-degree equations:

$$x + 3y = 5$$

$$x^2 = 9$$

$$3 + \sqrt{x} = 2(x - 7)$$

$$\frac{2}{x} = 7(x - 1)$$

$$(x - 3)(x + 3) = 12$$

Before proceeding, determine why each of the above is *not* a first-degree equation.

You should now be able to identify a first-degree equation and to distinguish it from equations that are not first degree. Remember that in any first-degree equation:

There is only one variable.
It is involved only in one or more of the four fundamental operations.
It is never multiplied by itself.
It is never in a denominator.

Now we are ready to begin solving first-degree equations. You probably remember the basic strategy—get all terms involving the variable on one side of the equal sign, and get all other terms on the other side.

To accomplish this, we use two rules:

Rule 1:   **A term may be transposed from one side of the equation to the other if and only if its sign is changed as it crosses the equal sign.**

EXAMPLE 1

$$\text{If} \quad x + 3 = 5,$$
$$\text{then} \quad x = 5 - 3$$
$$\text{and} \quad x = 2.$$

Note that the sign is changed.

EXAMPLE 2

$$\text{If} \quad 2 - x = 7,$$
$$\text{then} \quad 2 - 7 = x$$
$$\text{and} \quad -5 = x.$$

Rule 2:   **Both sides of an equation may be multiplied or divided by the same nonzero number.**

EXAMPLE 3

$$\text{If} \quad 3x = 2,$$
$$\text{then} \quad \frac{3x}{3} = \frac{2}{3},$$
$$\frac{\cancel{3}x}{\cancel{3}} = \frac{2}{3},$$
$$\text{and} \quad x = \frac{2}{3}.$$

EXAMPLE 4

$$\text{If} \quad \frac{x}{4} = 12,$$
$$\text{then} \quad 4\left(\frac{x}{4}\right) = 4(12),$$
$$\frac{\cancel{4}x}{\cancel{4}} = 48,$$
$$\text{and} \quad x = 48.$$

Now let's get on with the business of solving equations.

> Definition: A **solution** of an equation is any number that makes the equation true when that number is substituted for the variable. Sometimes it is called the **root** of the equation.

To solve first-degree equations, we will use a four-step procedure, which can be used to solve *all* first-degree equations.

The four steps are as follows:

> 1. **Simplify:**
>   a. Remove parentheses.
>   b. Multiply the entire equation by the common denominator of all fractions, if any.
>   c. Collect like terms.
>
> 2. **Transpose:**
>   Take all terms with variables to one side, and take everything else to the other side.
>
> 3. **Simplify.**
>
> 4. **Divide** the equation **by the coefficient** of the variable.

Here are some examples that illustrate the use of these steps in solving first-degree equations. Read through each step, and be sure you understand what has happened to the terms.

EXAMPLE 5

Solve the equation for x: $5x + 10 - 3x = 6 - 4x + 6$

Solution:

$$5x + 10 - 3x = 6 - 4x + 6$$

1. **Simplify:** collect like terms

$$2x + 10 = 12 - 4x$$

2. **Transpose.**

$$2x + 10 = 12 - 4x$$
$$2x + 4x = 12 - 10$$

3. **Simplify:** collect like terms.

$$6x = 2$$

4. **Divide by coefficient.**

$$\frac{6x}{6} = \frac{2}{6}$$

$$\frac{6x}{6} = \frac{1}{3}$$

So the answer is:

$$x = \frac{1}{3}$$

EXAMPLE 6

Solve the equation for x: $\dfrac{3x}{2} - 10 = 6$.

Solution:

$$\dfrac{3x}{2} - 10 = 6$$

1. **Simplify**: multiply.

$$2\left(\dfrac{3x}{2} - 10 = 6\right)$$

     (this cancels the denominator)

$$\dfrac{2 \cdot 3x}{2} - 2 \cdot 10 = 2 \cdot 6$$

     (cleared of the fraction)

$$3x - 20 = 12$$

2. **Transpose.**

$$3x - 20 = 12$$
$$3x = 12 + 20$$

3. **Simplify:** collect like terms.

$$3x = 32$$

4. **Divide by coefficient.**

$$\dfrac{3x}{3} = \dfrac{32}{3}$$

So the answer is:

$$x = \dfrac{32}{3}$$

EXAMPLE 7

Solve the equation for x: $3(x + 5) - 2x = \dfrac{x-1}{2} + 17$.

Solution:

$$3(x + 5) - 2x = \dfrac{x-1}{2} + 17$$

1. **Simplify:** remove parentheses,

$$3x + 15 - 2x = \dfrac{x-1}{2} + 17$$

     multiply,

$$2\left(3x + 15 - 2x = \dfrac{x-1}{2} + 17\right)$$

     (this cancels the denominator)

$$6x + 30 - 4x = \dfrac{2(x-1)}{2} + 34$$

     (cleared of fractions)

$$6x + 30 - 4x = x - 1 + 34$$

     collect like terms.

$$2x + 30 = x + 33$$

2. **Transpose.**

$$2x + 30 = x + 33$$

3. **Simplify:** collect like terms.

$$2x - x = 33 - 30$$

So the answer is:

$$x = 3$$

Try to solve this first-degree equation yourself. Cover the solution below, and refer to it only after you have arrived at *your* solution.

Problem 1

Solve the equation for x: $x - 3 - 2(6 - 2x) = 2(2x - 5).$

1. **Simplify:**   remove parentheses,   =
                    collect like terms.   =

2. **Transpose.**   =

3. **Simplify:**   collect like terms.   =

- -

Solution: $x - 3 - 2(6 - 2x) = 2(2x - 5)$

$$x - 3 - 12 + 4x = 4x - 10$$

$$5x - 15 = 4x - 10$$

$$5x - 4x = -10 + 15$$

$$x = 5$$

- -

Now try solving this equation without any clues.

Problem 2

Solve the equation for x: $3x - 2(x + 1) = 5x - 6.$

- -

Solution:

Answer: $x = 1$

- -

You should now be able to solve any first-degree equation. Remember that the sign of a term changes when the term is moved across the equal sign, and that both sides of an equation may be multiplied or divided by the same nonzero number.

Also remember the four basic steps:

1. **Simplify:** remove parentheses, clear of fractions, if any, collect like terms.
2. **Transpose.**
3. **Simplify.**
4. **Divide by coefficient.**

Before beginning the next unit you should solve the following equations.

EXERCISES

Solve:

1. $2x - 7 = 9 - 6x$

2. $2(x + 1) - 3(4x - 2) = 6x$

3. $\dfrac{x - 3}{4} = 5$

4. $20 - \dfrac{3x}{5} = x - 12$

5. $2x - 9 = 5x - 15$

6. $2(x + 2) = 5 + \dfrac{x + 1}{3}$

7. $15 - 3(9 - x) = x$

8. $3 - \dfrac{5(x - 1)}{2} = x$

9. $x - \dfrac{x - 1}{4} = 0$

10. $1 - \dfrac{x}{2} = 5$

If additional practice is needed:
Drooyan and Wooton, Unit 3, page 76, exercise 3
Peters, pags 156 and 157, problem 3a–e, i
Rees, Sparks, and Rees, page 74, problems 1–16
Rich, page 34, problem 30a–n

UNIT 4

Removing Multiple Grouping Symbols; Solving First-Degree Equations

In this unit we will consider several topics. You will learn to solve more difficult first-degree equations. Also, you will learn the difference between conditional equations and identities. When you have completed the unit, you will be able to solve first-degree equations with more than one level of grouped terms. You will also be able to distinguish between conditional equations and identities.

REMOVING MULTIPLE GROUPING SYMBOLS

Recall that in Unit 3 we learned four basic steps in solving first-degree equations:

> 1. **Simplify:** remove parentheses,
>   clear of fractions, if any,
>   collect like terms.
> 2. **Transpose.**
> 3. **Simplify.**
> 4. **Divide by coefficient.**

These same four steps apply when we solve equations of greater difficulty. For example, consider the equation:

$$2[2 - x - (2x - 5)] = 3 - x + 8$$

brackets ↑ ↑ parentheses

Note that one pair of symbols is "nested' within the outer pair. To simplify this equation, we start by removing the innermost pair and then work outward.

EXAMPLE 1

Solve this equation: $2[2 - x - (2x - 5)] = 3 - x + 8$.

- -

Solution: It will take two steps *just* to remove all the grouping symbols.

$$2[2 - x - (2x - 5)] = 3 - x + 8$$

$$2[2 - x - 2x + 5] = 3 - x + 8$$

$$4 - 2x - 4x + 10 = 3 - x + 8$$

With the symbols removed, the solution continues as in Unit 3:

$$-6x + 14 = 11 - x$$
$$-6x + x = 11 - 14$$
$$-5x = -3$$
$$\frac{-5x}{-5} = \frac{-3}{-5}$$
$$x = \frac{3}{5}$$

- -

Try solving this problem yourself.

Problem 1

Solve this equation: $3\{1 + 4x - [x + 1]\} = 0$.

- -

Solution:

Answer: $x = 0$

- -

TYPES OF EQUATIONS

So far, all the equations we have considered have been what are called conditional equations. A **conditional equation** is one that is true for only certain values of a variable.

For example, $x + 5 = 8$ is true only for $x = 3$. Therefore $x + 5 = 8$ is a conditional equation.

Consider also this example.

EXAMPLE 2

Solve this equation: $2x + 3 = 7$.

- -

Solution: $2x + 3 = 7$

$$2x = 7 - 3$$

$$\frac{2x}{2} = \frac{4}{2}$$

$$x = 2$$

Thus $x = 2$ is the solution of $2x + 3 = 7$, and $2x + 3 = 7$ is a conditional equation. It is true only when $x = 2$. If we substitute $x = 2$ into the equation, we obtain

$$2(2) + 3 = 7$$

$$4 + 3 = 7$$

$$7 = 7$$

which is true. So we say $x = 2$ satisfies the equation.

- -

Now, in contrast, consider the following equation.

EXAMPLE 3

Solve this equation: $2x + 1 + x = 3x + 1$

Solution:
$$2x + 1 + x = 3x + 1$$
$$3x + 1 = 3x + 1$$
$$3x - 3x = 1 - 1$$
$$0 = 0$$

The solution is the entire set of real numbers.

This equation is *always* true, regardless of the value of x, because 0 always equals 0. We call this type of equation an **identity**. Its solution is the entire set of reals. In other words, x can equal any number.

For example, if $x = 1$, then $2(1) + 1 + (1) = 3(1) + 1$

and $2 + 2 = 3 + 1$.

Or,   if $x = 7$, then $2(7) + 1 + (7) = 3(7) + 1$

$$14 + 8 = 21 + 1$$

and $22 = 22$.

Try *any* value you like for x. Prove to yourself that it does indeed satisfy the equation. For instance, try $x = -3$.

> Definition: An equation that is true for only certain values is called a **conditional equation**.

> Definition: An equation that is true for all values of the variable is called an **identity**.

Thus, an equation that is true for *all* values of the variable is called an identity. Convince yourself that the following equation is an identity.

Problem 2

Question: Is this equation an identity? $3x + 4(x - 2) = x - 5 + 3(2x - 1)$

- -

Solution:

Answer: yes

- -

In summary, we have learned that the four basic steps—simplify, transpose, simplify, and divide—can be used to solve even more complicated first-degree equations and that, when several different grouping symbols are used, the innermost pairs are removed first. We also distinguished between **conditional equations**, which are true only for certain values of the variable, and **identities**, which are true for any real value of the variable.

Now try to solve the following equations. In addition, identify each as a conditional equation or an identity.

EXERCISES

Solve:

1. $-2[x - (4x - 5)] = 5x$

2. $1 - 2[4 - (3x - 5)] = 3x + 1$

3. $1 - \{x - (1 - x)\} = 0$

4. $2\{x + 4 - 3(2x - 1)\} = 3(4 - 3x) + 2 - x$

5. $\{[x - 2(x + 1) + 2] + 4\} = 4$

6. $12x + 4 - [(3 - x) - (5x + 7)] = 13$

7. $2 - 3\{2x - 2[1 - (2x + 1)] + x\} = 23$

8. $x - \dfrac{x+2}{2} = 3x - (1 + 2x) - \dfrac{x}{2}$

9. $2x - \{1 - [x - (3 - 2x)]\} = 5x - 4$

10. $20 - \{[3x - (x - 1)] - 5x\} = 0$

If additional practice is needed:
   Leithold, page 143, problems 7–12
   Lial and Miller, page 65, problems 3, 4

UNIT 5

A Special Type of Equation: The Fractional Equation

In this unit you will learn about a special type of equation, the fractional equation. When you have completed the unit, you will be able to identify and solve fractional equations.

> Definition: A **fractional equation** is an equation in which the variable appears in a denominator.

For example,

$$\frac{2+x}{x} = 3$$

is a fractional equation.

The same four steps we have been using will again be used to solve fractional equations. However, we must now add a fifth step in which we check to see whether our solution satisfies the **original** equation we are solving.

There are several operations that may produce an equation not equivalent to the original equation. One of these operations is multiplying or dividing both sides of an equation by an expression containing the variable. The final test of whether a number is part of the solution is to insert it into the original equation and see whether it yields a true statement.

Recall the four steps:

1.	**Simplify:** remove parentheses, clear of fractions, if any, collect like terms.
2.	**Transpose.**
3.	**Simplify.**
4.	**Divide by coefficient.**
5.	**Check** by substituting the tentative answer in the original equation.

Now we must also:

At this time it is necessary to remind you that **division by zero is undefined**. In other words, zero can never be the denominator of a fraction; $\frac{a}{0}$, for example, is meaningless.

We will now use the four basic steps and the check by substitution to solve our example of a fractional equation.

EXAMPLE 1

Solve for x: $\dfrac{2+x}{x} = 3$.

Solution: 1. **Simplify:** clear of fractions.

$$\left(\frac{2+x}{x} = 3\right) x$$

$$\frac{(2+x)}{x} \cdot x = 3 \cdot x$$

$$\frac{(2+x)}{\not{x}} \cdot \not{x} = 3 \cdot x$$

$$2 + x = 3x$$

2. **Transpose.**

$$2 = 3x - x$$

3. **Simplify.**

$$2 = 2x$$

4. **Divide** by coefficient.

$$\frac{\not{2}}{\not{2}} = \frac{\not{2}x}{\not{2}}$$

$$1 = x$$

5. **Check:**   substitute the tentative answer in the **original** equation.

$$\frac{2+x}{x} = 3$$

$$\frac{2+1}{1} \overset{?}{=} 3$$

$$3 = 3$$

This is true. So $x = 1$ is the solution.

EXAMPLE 2

Solve for x: $\dfrac{5}{x} + 5 = \dfrac{-1-x}{2x}$.

Solution: What shall we use as the common denominator to solve the given equation:

$$x, \quad 2x, \quad 2x^2, \quad 3x, \quad \text{etc.?}$$

Using our method, we can make a wrong choice and still solve the equation simply by adding an extra step or two.

Suppose we use x:

$$\frac{5}{x} + 5 = \frac{-1-x}{2x}$$

$$x\left(\frac{5}{x} + 5 = \frac{-1-x}{2x}\right)$$

$$\frac{x \cdot 5}{x} + x \cdot 5 = \frac{-1-x}{2x} \cdot x$$

$$5 + 5x = \frac{-1-x}{2}$$

Obviously x was not a wise choice as it did not clear the equation completely of fractions. However, it is easy to continue the solution.

Because there is still a denominator of 2, we can simply repeat the process using 2:

$$\left(5 + 5x = \frac{-1-x}{2}\right) \cdot 2$$

$$5 \cdot 2 + 5x \cdot 2 = \frac{(-1-x)\,2}{2}$$

$$10 + 10x = -1 - x$$

$$10x + x = -1 - 10$$

$$11x = -11$$

$$\frac{11x}{11} = \frac{-11}{11}$$

$$x = -1$$

Had we multiplied by $2x$ in the first step, the problem would have been much shorter.

Now we must check the tentative solution, using the original equation.

$$\frac{5}{x} + 5 = \frac{-1-x}{2x}$$

$$\frac{5}{-1} + 5 \stackrel{?}{=} \frac{-1-(-1)}{2(-1)}$$

$$-5 + 5 \stackrel{?}{=} \frac{-1+1}{-2}$$

$$0 \stackrel{?}{=} \frac{0}{-2}$$

$$0 = 0$$

This is true. So $x = -1$ is the solution.

Now you try one.

Problem 1

Solve for x: $\dfrac{2}{x + 1} + 3 = 1$.

Hint: $(x + 1)$ is the **entire** denominator.

Solution:

Answer: $x = -2$

Here are two more examples.

EXAMPLE 3

Solve for x: $\dfrac{2(x + 1)}{x} = \dfrac{2}{x}$.

Solution:

$$\dfrac{2(x + 1)}{x} = \dfrac{2}{x}$$

$$\left(\dfrac{2x + 2}{x} = \dfrac{2}{x}\right) x$$

$$\left(\dfrac{2x + 2}{x}\right) x = \dfrac{2}{x} \cdot x$$

$$2x + 2 = 2$$

$$2x = 2 - 2$$

$$\dfrac{2x}{2} = \dfrac{0}{2}$$

$$x = 0$$

Check: $\dfrac{2(x + 1)}{x} = \dfrac{2}{x}$

$$\dfrac{2(0 + 1)}{0} \overset{?}{=} \dfrac{2}{0}$$

However, recall that we said that **division by zero is undefined**! Therefore $x = 0$ cannot be a solution to this equation.

Answer: The equation has no solution.

What this example should impress upon you is the necessity for checking the purported solution.

EXAMPLE 4

Solve for x: $\dfrac{2x - 4}{x - 3} = 3 + \dfrac{2}{x - 3}$.

Solution:
$$\frac{2x - 4}{x - 3} = 3 + \frac{2}{x - 3}$$

$$\frac{2x - 4}{x - 3}(x - 3) = 3(x - 3) + \frac{2}{x - 3}(x - 3)$$

$$2x - 4 = 3(x - 3) + 2$$

$$2x - 4 = 3x - 9 + 2$$

$$2x - 4 = 3x - 7$$

$$2x - 3x = -7 + 4$$

$$-x = -3$$

$$x = 3$$

Check:
$$\frac{2x - 4}{x - 3} = 3 + \frac{2}{x - 3}$$

$$\frac{6 - 4}{3 - 3} \stackrel{?}{=} 3 + \frac{2}{3 - 3}$$

But division by zero is undefined, so 3 cannot be a solution.

Answer: The equation has no solution.

It's your turn again.

Problem 2

Solve for x: $\dfrac{3}{2x - 1} = 5.$

Solution:

Answer: $x = \frac{4}{5}$

Fractional equations of the following type occur quite often:

$$\frac{2}{x + 1} = \frac{3}{x}$$

There are **two** fractions in this equation. One is on each side of the equal sign.

There is an easy way to solve this kind of fractional equation: that is to "cross-multiply."

$$\frac{2}{x + 1} \diagdown\kern-1em\diagup \frac{3}{x}$$

$$2x = 3(x + 1)$$

What we actually did was to multiply the entire equation by the lowest common denominator, $x(x + 1)$. But by simply "cross-multiplying" we save ourselves a few steps. (If you don't believe me, try it the long way.) Then the problem continues on as before.

$$2x = 3x + 3$$

$$2x - 3x = 3$$

$$-x = 3$$

$$x = -3$$

Check: $\qquad \dfrac{2}{x + 1} = \dfrac{3}{x}$

$$\frac{2}{(-3) + 1} \overset{?}{=} \frac{3}{(-3)}$$

$$\frac{2}{-2} \overset{?}{=} \frac{3}{-3}$$

$$-1 = -1$$

This is true. So $x = -3$ is the solution.

EXAMPLE 5

Solve for x: $\quad \dfrac{5}{2x - 1} = \dfrac{3}{x + 1}$.

- -

Solution: $\qquad \dfrac{5}{2x - 1} \diagdown\kern-1em\diagup \dfrac{3}{x + 1}$

$$5(x + 1) = 3(2x - 1)$$

$$5x + 5 = 6x - 3$$

$$5x - 6x = -3 - 5$$

$$-x = -8$$

$$x = 8$$

Check: $$\frac{5}{2x-1} = \frac{3}{x+1}$$

$$\frac{5}{2(8)-1} \overset{?}{=} \frac{3}{(8)+1}$$

$$\frac{1}{3} = \frac{1}{3}$$

This is true. So $x = 8$ is the solution.

_____

EXAMPLE 6

Solve for x: $\dfrac{3x}{2x-3} = 4$.

_____

Solution: Since a whole number can always be divided by 1 without changing its value,

we can write 4 as $\dfrac{4}{1}$ and then cross-multiply.

$$\frac{3x}{2x-3} = \frac{4}{1}$$

$$3x(1) = 4(2x-3)$$

What would have happened if we chose to multiply through by the lowest common denominator instead? Finish the problem yourself.

Answer: $x = \frac{12}{5}$

_____

   You should now be able to identify and solve fractional equations. Remember that an equation of this type has a variable in a denominator. In fact, there may be a fraction on either side or both sides of the equal sign.

   The same four basic steps—simplify, transpose, simplify, and divide—are used to solve fractional equations. In addition, we must check the tentative solution by substituting it in the original equation.

   Remember also that, if we make a wrong choice for the common denominator and a denominator remains after completing the basic steps, we can simply repeat the process, using the remaining denominator.

   Finally, remember that, when there is a **single** fraction on both sides of the equal sign, we can use a shortcut: cross-multiplying the fractions.

   Before beginning the next unit you should solve the following equations.

EXERCISES

Solve for x:

1. $1 = \dfrac{5}{x}$

2. $\dfrac{x-3}{2} = \dfrac{2x+4}{5}$

3. $\dfrac{6}{x-2} = -3$

4. $\dfrac{3x-3}{x-1} = 2$

5. $\dfrac{x}{2} = \dfrac{x+6}{5}$

6. $\dfrac{3}{x} = \dfrac{4}{x-2}$

7. $\dfrac{4}{x+3} = \dfrac{1}{x-3}$

8. $\dfrac{5-2x}{x-1} = -2$

9. $\dfrac{x+3}{x-2} = 2$

10. $\dfrac{2}{x} + \dfrac{3(x-1)}{5x} = 1$

11. $5 + \dfrac{3+x}{x} = \dfrac{5}{x}$

12. $\dfrac{4}{x-2} - \dfrac{1}{x} = \dfrac{5}{x-2}$

If additional practice is needed:
  Lial and Miller, page 65, problems 5–11
  Peters, page 157, problem 3j, k, n
  Rees, Sparks, and Rees, page 74, problems 21–28
  Rich, page 83, problems 28–34

UNIT 6

Another Special Type of Equation: The Literal Equation

In this unit you will learn about another special type of equation, the literal equation. When you have completed the unit, you will be able to identify and solve literal equations.

> Definition: A **literal equation** is an equation that contains letters, in addition to a variable and numbers.

We therefore solve a literal equation for the stated variable in terms of these other letters. Again, the four basic steps we have been using for other first-degree equations may be used for solving literal equations. Recall that the four steps are:

> 1. **Simplify:** remove parentheses, clear of fractions, if any, collect like terms.
> 2. **Transpose.**
> 3. **Simplify.**
> 4. **Divide by coefficient.**

This is an example of a literal equation:

$$2x - 4p = 3x + 2p$$

Although literal equations may appear at first glance to be more difficult, you will find that they are easier to solve than many other equations you have done.

EXAMPLE 1

Solve this literal equation for x: $2x - 4p = 3x + 2p$.

Solution: $2x - 4p = 3x + 2p$

$$2x - 3x = 2p + 4p$$

$$-x = 6p$$

$$\frac{-x}{-1} = \frac{6p}{-1}$$

$$x = -6p$$

Now *you* solve the same equation for p.

Problem 1

Solve this literal equation for p: $2x - 4p = 3x + 2p$.

Solution:

Answer: $p = \dfrac{-x}{6}$

Let's do two more examples.

EXAMPLE 2

Solve this equation for x: $ax - 3 = -2cx$

Solution: $ax - 3 = -2cx$

$$ax - 3 = -2cx$$

Transpose: $ax + 2cx = 3$

Recall from Unit 2 the distributive property, $a(b + c) = ab + ac$, which we used to remove parentheses. Now, in order to find the coefficient of x, we will use the distributive property in reverse.

Notice that x is common to both terms on the left side of the equation. Using the distributive property, we can factor out the common x since

$$ax + 2cx = x(a + 2c)$$

Factor out x: $x(a + 2c) = 3$

$$\frac{x(a + 2c)}{a + 2c} = \frac{3}{a + 2c}$$

$$x = \frac{3}{a + 2c}$$

Notice that for our literal equations we do *not* get a nice simple number for an answer; instead we get a fairly complicated looking expression for x in terms of the other letters.

EXAMPLE 3

Solve this equation for x: $3x + 5y = ax + 2y.$

Solution: $3x + 5y = ax + 2y$

$$3x - ax = 2y - 5y$$

$$3x - ax = -3y$$

Factor out x. $x(3 - a) = -3y$

$$\frac{x(3 - a)}{3 - a} = \frac{-3y}{3 - a}$$

$$x = \frac{-3y}{3 - a}$$

Or, if you prefer,
multiply top and bottom $x = \frac{3y}{a - 3}$
by -1.

More often than not we solve equations for y in terms of x. Using the same equation as in Example 3, try it.

Problem 2

Solve this equation for y: $3x + 5y = ax + 2y$.

Solution:

Answer: $y = \dfrac{ax - 3x}{3}$

We'll do two more examples.

EXAMPLE 4

Solve this equation for y: $\dfrac{3y + a}{a} = \dfrac{4y + b}{b}$.

Solution: $\dfrac{3y + a}{a} = \dfrac{4y + b}{b}$ Shortcut: cross-multiply from Unit 5.

$$b(3y + a) = a(4y + b)$$
$$3by + ab = 4ay + ab$$
$$3by - 4ay = ab - ab$$
$$y(3b - 4a) = 0$$
$$\frac{y(3b - 4a)}{(3b - 4a)} = \frac{0}{3b - 4a}$$
$$y = 0$$

QUESTION: Must we now check to see whether this answer is indeed the solution to the equation? In other words, is this a fractional equation?

EXAMPLE 5

Solve for x: $a(x-1) = -\dfrac{x}{b}$.

Solution:

$$a(x-1) = -\frac{x}{b}$$

$$ax - a = -\frac{x}{b}$$

$$\left(ax - a = -\frac{x}{b}\right)b$$

$$abx - ab = -x$$

$$abx + x = ab$$

Factor out x: $\quad x(ab+1) = ab$

$$\frac{x(\cancel{ab+1})}{(\cancel{ab+1})} = \frac{ab}{ab+1}$$

$$x = \frac{ab}{ab+1}$$

Now it's your turn to solve the above equation for b.

Problem 3

Solve for b: $a(x-1) = -\dfrac{x}{b}$.

Solution:

Answer: $b = \dfrac{-x}{a(x-1)}$

You should now be able to identify and solve any literal equation. Remember that a literal equation contains letters in addition to the variable and numbers. Even though literal equations are a special type of equation, the same four basic steps are used to solve them.

Remember also that, as we learned in Unit 5, when there is a **single** fraction on both sides of the literal equation we can cross-multiply.

Before beginning the next unit you should solve the following literal equations.

EXERCISES

Solve for a:

1. $\dfrac{2ax}{3c} = \dfrac{y}{m}$

2. $2cy + 4d = 3ax - 4b$

3. $ax + 3a = bx + 7c$

4. $3(x - a) = 2a - x - \dfrac{b - x}{c}$

5. $a(x + 2) = \pi - cy$

6–10. Now solve each of the above equations for x.

11–15. Now solve each of the above equations for c.

If additional practice is needed:
Drooyan and Wooton, Unit 3, page 76, exercise 4
Leithold, page 143, problems 29–32
Peters, page 160, problem 1a–e
Rees, Sparks, and Rees, page 75, problems 33–36

UNIT 7

Applied Problems

In this unit we will consider several applied problems. First you will learn how to translate the applied problem into a mathematical equation. This equation can then be solved using the methods you have learned in Units 2–6.

CHANGING VERBAL STATEMENTS TO ALGEBRAIC EXPRESSIONS

In algebra, as you know, letters are used to represent numbers. By using letters and mathematical symbols, we can replace lengthy verbal statements with short algebraic expressions. Here are a few examples.

EXAMPLE 1	The sum of a number and 7	$x + 7$
EXAMPLE 2	Five minus some number	$5 - y$
EXAMPLE 3	Three times a number plus 2	$3w + 2$
EXAMPLE 4	A number divided by 11 is 20.	$\dfrac{z}{11} = 20$

You try the next few.

Let x represent a number. Write each of the following in terms of x.

Problem 1   Three times a number minus 2

Problem 2   Seven divided by a number

Problem 3   Five plus a number

Answers: 1. $3x - 2$ 2. $\dfrac{7}{x}$ 3. $5 + x$

CHANGING VERBAL STATEMENTS TO ALGEBRAIC EQUATIONS

We'll continue translating words into symbols, but now we'll include an equal sign.

EXAMPLE 5   Three times a number minus 2 equals 10.

Solution: $3x - 2 = 10$

EXAMPLE 6   Five times a number plus 2 is 10

Solution: $5x \qquad\qquad + \ 2 = 10$

Here are three for you to try. For each problem, write an equation and then solve it.

Problem 4   Five plus a number is 7.

Answer: $5 + x = 7$

Problem 5   When a number is decreased by 3, the result is 15.

Answer: $x - 3 = 15$

- -

Problem 6   The sum of a number and 11 is 12.

Answer: $x + 11 = 12$

- -

Before trying to solve applied problems, we need to try a few more translations.

EXAMPLE 7

The sum of two numbers is 12. If x represents one of these numbers, express the other in terms of x.

- -

Solution: Let x represent one number.

Let $\triangle$ represent the other number.

The sum of the two numbers is 12.

$$x \quad + \qquad\qquad \triangle \qquad = 12$$
$$\triangle \qquad = 12 - x$$

Thus the other number   is $12 - x$.

- -

EXAMPLE 8

The sum of two numbers is 5. If x represents one of these numbers, express the other in terms of x.

- -

Solution: The other number is $5 - x$.

- -

I think we're ready to look at some applied problems.

SOLVING APPLIED PROBLEMS

We'll first translate the problem into algebraic symbols, then solve the equation, and finally answer the original question.

EXAMPLE 9

Bob is 5 years older than Barbara. The sum of their ages is 23. How old is Barbara?

Solution: Let x represent Barbara's age.

Then Bob's age is $x + 5$ because he is 5 years older.

The sum of their ages is 23.

Bob's age + Barbara's age = 23

$$(x + 5) \quad + \qquad\qquad x = 23$$
$$2x + 5 = 23$$
$$2x = 18$$
$$x = 9$$

Barbara is 9 years old.

The next one is a bit longer, so we'll do it in sections.

EXAMPLE 10

In the election for homecoming queen, Sally received 3 more votes than Dawn and 2 fewer than Joy.

If Sally received x votes:

a.   How many votes did Dawn receive?

b.   How many votes did Joy receive?

c.   If a total of 44 votes were cast for the three girls, how many votes did each receive?

Solution: a.   If Sally received x votes, and that number was 3 more than Dawn received, then Dawn received $x - 3$.

b.   If Sally received 2 fewer votes than Joy, then Joy received 2 votes more than Sally, or $x + 2$.

c. $x + (x - 3) + (x + 2) = 44$
      ↑   ↑   ↑
   Sally Dawn   Joy

$$3x - 1 = 44$$
$$3x = 45$$
$$x = 15$$

Sally received 15 votes.
Dawn received 15 − 3, or 12 votes.
Joy received 15 + 2, or 17 votes.

_____

I think we've done enough examples.

By now you should be able to translate verbal expressions into algebraic symbols, and to translate an applied problem into a mathematical equation.

Before going on to the next unit, do the following exercises.

EXERCISES

Let x represent a number. Express each of the following in terms of x:

1.   A number decreased by 5.
2.   Three times a number increased by 8.
3.   Eight times a number minus 10.
4.   A number divided by 3.

Let x represent a number. Express each of the following as an equation:

5.   Two times a number decreased by 5 equals 11.
6.   Seven times a number is 35.
7.   A number added to 20 is the same as 32.
8.   The sum of x and 12 is 20.
9.   Fifteen increased by 2 times a number is 47.
10.   Four more than three times a number is 17.

Solve each of the following:

11.   George is 8 years older than Jack. The sum of their ages is 42. How old is each person?

12.   A rope that is 36 feet in length is cut into two pieces. If one piece is 10 feet longer than the other, how long is each piece?

_____

UNIT 8

Positive Integral Exponents

In this unit you will learn to simplify expressions involving positive integral exponents. You will also learn four of the **five basic laws of exponents**. When you have completed the unit, you will be able to simplify expressions in which there is an exponent that is a positive integer or 0.

RECOGNIZING EXPONENTS

Consider the expression b^n. This is read as "b to the nth power." The b is referred to as the **base**, and the n is the **exponent**. Here $b \neq 0$.

In this unit we will consider exponents that are 0 or a positive integer (1, 2, 3, 4, etc.) only. Negative and fractional exponents will be discussed in Units 9 and 13.

Definition: **Positive Integral Exponent**

$$b^n = \underbrace{b \cdot b \cdot b \cdot \; \cdots \; \cdot b}_{n \text{ factors}}, \text{ if } n \text{ is a positive integer}$$

Consider the positive integral exponents in these expressions.

EXAMPLE 1 $b^2 = b \cdot b$

EXAMPLE 2 $x^3 = x \cdot x \cdot x$

EXAMPLE 3 $2^5 = 2 \cdot 2 \cdot 2 \cdot 2 \cdot 2 = 32$

Note that, as in preceding units, if no exponent is written, the exponent is understood to be 1.

> Definition: **Zero Exponent**
> $$b^0 = 1$$

EXAMPLE 4 $x^0 = 1$

EXAMPLE 5 $5^0 = 1$

EXAMPLE 6 $(3ab + \pi - 5\sqrt{7})^0 = 1$

SIMPLIFYING EXPRESSIONS WITH EXPONENTS

To accomplish simplification, we have five basic laws of exponents. We will discuss only four at this time.

These laws are used to shorten our work. When in doubt about any of these laws, we can always go back to the definitions on the first and second pages of this unit.

> Laws of Exponents
>
> I. **Multiplication** $b^n \cdot b^m = b^{n+m}$
>
> II. **Power of a power** $(b^n)^m = b^{nm}$
>
> III. **Power of a product** $(bc)^n = b^n c^n$
>
> IV. **Power of a fraction** $\left(\dfrac{a}{b}\right)^n = \dfrac{a^n}{b^n}$

> I. **Multiplication** $b^n \cdot b^m = b^{n+m}$

EXAMPLE 7

Simplify: $a^2 \cdot a^3$.

Solution: $a^2 \cdot a^3 = a^{2+3} = a^5$

If you doubt this, use the definition: $a^2 \cdot a^3 = (a \cdot a) \cdot (a \cdot a \cdot a) = a^5$.

EXAMPLE 8

Simplify: $x \cdot x^2$.

Solution: $x \cdot x^2 = x^1 \cdot x^2 = x^{1+2} = x^3$

EXAMPLE 9

Simplify: $5 \cdot 5^2$.

Solution: $5 \cdot 5^2 = 5^1 \cdot 5^2 = 5^{1+2} = 5^3$

In case you're not convinced:

$$5 \cdot 5^2 = 5 \cdot (5 \cdot 5) = 5^3$$

EXAMPLE 10

Simplify: $a^2(2a^3)$.

Solution: $a^2(2a^3) = a^2 \cdot 2 \cdot a^3 = 2a^5$

EXAMPLE 11

Simplify: $(a^2b^3)(ab^4)(3abc)$.

Solution: $(a^2b^3)(ab^4)(3abc)$

$a^2 \cdot b^3 \cdot a^1 \cdot b^4 \cdot 3 \cdot a^1 \cdot b^1 \cdot c^1 = 3a^{2+1+1}b^{3+4+1}c^1 = 3a^4b^8c$

You try some.

Problem 1 $x^3 \cdot x^5 =$

Problem 2 $x^2(x^3y) =$

Problem 3 $(x^2y^3)(x^7y) =$

Answers: 1. x^8 2. x^5y 3. x^9y^4

| II.   **Power of a power** $(b^n)^m = b^{nm}$ |

EXAMPLE 12

Simplify: $(a^2)^3$.

Solution: $(a^2)^3 = a^{2 \cdot 3} = a^6$

If we had used the definition instead:

$$(a^2)^3 = (a^2)(a^2)(a^2)$$
$$= a^{2+2+2}$$
$$= a^6$$

Note that the laws of exponents are shortcut methods. With them, you do not have to work out expressions completely with the definition. But they are shortcuts only if you apply them properly.

EXAMPLE 13

Simplify: $(x^{15})^2$.

Solution: $(x^{15})^2 = x^{15 \cdot 2} = x^{30}$

Now it's your turn again.

Problem 4 $(a^{10})^2 =$

Problem 5 $(x^3)^0 =$

Problem 6 $c^{10} \cdot c^2 =$

Answers: 4. a^{20} 5. 1   6. c^{12}

$$\boxed{\text{III.} \quad \textbf{Power of a product} \qquad (bc)^n = b^n c^n}$$

Notice that this law is really just an extension of Law II: $(b^n)^m = b^{nm}$.

EXAMPLE 14

Simplify: $(3x^2)^2$.

Solution: $(3x^2)^2 = 3^2 x^4 = 9x^4$

Be careful; most people forget that the 3 is to be squared too.

EXAMPLE 15

Simplify: $(x^2 y^3)^4$.

Solution: $(x^2 y^3)^4 = x^8 y^{12}$

$$\boxed{\text{IV.} \quad \textbf{Power of a fraction} \qquad \left(\frac{a}{b}\right)^n = \frac{a^n}{b^n}}$$

EXAMPLE 16

Simplify: $\left(\dfrac{2}{5}\right)^2$

Solution: $\left(\dfrac{2}{5}\right)^2 = \dfrac{2^2}{5^2} = \dfrac{4}{25}$

EXAMPLE 17

Simplify: $\left(\dfrac{x^2}{c^3}\right)^5$.

Solution: $\left(\dfrac{x^2}{c^3}\right)^5 = \dfrac{x^{2\cdot5}}{c^{3\cdot5}} = \dfrac{x^{10}}{c^{15}}$

Have you noticed that Laws II, III, and IV deal with removing parentheses? To remove the parentheses we multiply the exponents.

Sound familiar? To remove parentheses, we multiply. . . .

We are now ready to combine the laws in a single problem.

Laws of Exponents		
I.	**Multiplication**	$b^n \cdot b^m = b^{n+m}$
II.	**Power of a power**	$(b^n)^m = b^{nm}$
III.	**Power of a product**	$(bc)^n = b^n c^n$
IV.	**Power of a fraction**	$\left(\dfrac{a}{b}\right)^n = \dfrac{a^n}{b^n}$

My choice of procedure will be to remove the parentheses first and finish simplifying the expression by using Law I.

EXAMPLE 18

Simplify: $a^2(a^3)^2$.

Solution: $a^2(a^3)^2 = a^2 \cdot a^6 = a^{2+6} = a^8$

As in solving equations, I removed the parentheses first.

EXAMPLE 19

Simplify: $x^3(3\pi x)^0$.

Solution: $x^3(3\pi x)^0 = x^3 \cdot 1 = x^3$

Try another yourself.

Problem 7

Simplify: $x^3(xy^2)^2$.

- -

Solution:

Answer: x^5y^4

- -

You should now be able to simplify any expression in which an exponent of 0 or some positive integer appears. Remember that, when you are in doubt about one of the laws of exponents, you can always return to the definition of a positive integral exponent or a zero exponent.

Now try to simplify the following expressions.

EXERCISES

Simplify:

1. $(3y)^2 \cdot (2y)^3$

2. $3x^0$

3. $x^2(x^3)^4$

4. $\left(\dfrac{a^2b^3cd^5}{3x^2w^0} \right)^7$

5. $\dfrac{(2ab)^2}{(3x^3)^2}$

6. $(3x^5)^2(2x^3)^3$

7. $(x^2y)(xy^2)$

8. $2(3ab^2)^2$

9. $(-4c)^2$

10. $\left(\dfrac{xyz^2}{5a}\right)^3$

11. $(-2abc)(bcd)(3abc^2)$

12. $(2x^2yz)(-5xz)^2(xyz^2)^3$

If additional practice is needed:
    Lial and Miller, page 28, problems 41–46
    Peters, page 84, problems 1–13, 18–26
    Rich, page 68, problems 21–23

UNIT 9

Negative Exponents

In this unit you will learn how to handle **negative exponents**. When you have completed the unit, you will be able to simplify expressions with negative exponents.

> **Definition:** **Negative exponent**
>
> $$b^{-n} = \frac{1}{b^n} \quad \text{and} \quad \frac{1}{b^{-n}} = b^n$$

since

$$\frac{1}{b^{-n}} = \frac{1}{\dfrac{1}{b^n}} = 1 \div \frac{1}{b^n} = 1 \cdot b^n = b^n$$

Therefore a negative exponent has the effect of moving a **factor** from top to bottom (or vice versa).

Consider Examples 1–11, and be certain you understand the simplification.

EXAMPLE 1 $\quad 2^{-1} = \frac{1}{2}$

EXAMPLE 2 $\quad a^{-1} = \dfrac{1}{a}$

EXAMPLE 3 $\quad b^{-2} = \dfrac{1}{b^2}$

EXAMPLE 4 $\quad c^{-3} = \dfrac{1}{c^3}$

EXAMPLE 5 $\quad 2x^{-1} = \dfrac{2}{x}$

EXAMPLE 6 $\quad 3ab^{-2} = \dfrac{3a}{b^2}$

EXAMPLE 7 $5^{-1}ab^{-2}c = \dfrac{ac}{5b^2}$

EXAMPLE 8 $\dfrac{1}{a^{-2}} = a^2$

EXAMPLE 9 $\dfrac{5}{x^{-3}y} = \dfrac{5x^3}{y}$

EXAMPLE 10 $\dfrac{a^{-2}}{b^{-3}} = \dfrac{b^3}{a^2}$

EXAMPLE 11 $\dfrac{y^{-3}}{x^2} = \dfrac{1}{y^3x^2}$

When simplifying expressions dealing with exponents, the objective is to write the final answer *without* zero or negative exponents. Luckily the laws of exponents we introduced in Unit 8 apply for all types of exponents. We can use these laws and our definition to simplify expressions with negative exponents.

Laws of Exponents		
I.	**Multiplication**	$b^n \cdot b^m = b^{n+m}$
II.	**Power of a power**	$(b^n)^m = b^{nm}$
III.	**Power of a product**	$(bc)^n = b^n c^n$
IV.	**Power of a fraction**	$\left(\dfrac{a}{b}\right)^n = \dfrac{a^n}{b^n}$

EXAMPLE 12

Simplify: $(10^{-3})^2$.

There are many possible ways to simplify such expressions. I will attempt to use one technique for all:

Remove the parentheses.
Write with positive exponents.
Use the **multiplication** law of exponents.

Solution: $(10^{-3})^2 = 10^{-6} = \dfrac{1}{10^6}$

EXAMPLE 13

Simplify: $(xy^{-1})^{-3}$.

Solution: $(xy^{-1})^{-3} = x^{-3}y^3 = \dfrac{y^3}{x^3}$

EXAMPLE 14

Simplify: $\dfrac{ab^{-4}}{a^{-2}b}$.

Solution: $\dfrac{ab^{-4}}{a^{-2}b} = \dfrac{a \cdot a^2}{b^4 \cdot b} = \dfrac{a^3}{b^5}$

EXAMPLE 15

Simplify: $\left(\dfrac{x^2}{y^{-3}w}\right)^{-1}$

Solution: $\left(\dfrac{x^2}{y^{-3}w}\right)^{-1} = \dfrac{x^{-2}}{y^3w^{-1}} = \dfrac{w}{x^2y^3}$

Now try two yourself.

Problem 1

Simplify: $\left(\dfrac{x^2y^0}{w^{-1}}\right)^{-2}$

Solution:

Answer: $\dfrac{1}{w^2x^4}$

Problem 2

Simplify: $\left(\dfrac{a}{b}\right)^{-1}$.

Solution:

Notice what happened—the fraction is simply inverted. *You have just proved a theorem!*

Hence

$$\left(\frac{a}{b}\right)^{-1} = \frac{b}{a}$$

We'll end the unit with three more examples.

EXAMPLE 16 $\qquad \left(\dfrac{3}{4}\right)^{-1} = \dfrac{4}{3}$

EXAMPLE 17 $\qquad \left(\dfrac{a^2b}{c^3}\right)^{-1} = \dfrac{c^3}{a^2b}$

Try the next one yourself before looking at the answer.

EXAMPLE 18

Simplify: $\left[\dfrac{b^2}{(a^2b)^{-2}}\right]^{-1}$

Solution:

| Method 1 | Method 2 |

$$\left[\frac{b^2}{(a^2b)^{-2}}\right]^{-1} = \frac{(a^2b)^{-2}}{b^2}$$

$$\left[\frac{b^2}{(a^2b)^{-2}}\right]^{-1} = \left[\frac{b^2}{a^{-4}b^{-2}}\right]^{-1}$$

$$= \frac{a^{-4}b^{-2}}{b^2}$$

$$= \frac{b^{-2}}{a^4b^2}$$

$$= \frac{1}{a^4b^2b^2}$$

$$= \frac{1}{a^4b^2b^2}$$

$$= \frac{1}{a^4b^4}$$

$$= \frac{1}{a^4b^4}$$

You should now be able to simplify expressions involving negative exponents. Try simplifying the expressions in the exercises, that is, write them without negative exponents or parentheses.

EXERCISES

Simplify:

1. $\dfrac{a^{-3}}{a^2}$

2. $\dfrac{a^{-2}x^3}{y^{-1}}$

3. $(x^2y)^{-2}$

4. $\dfrac{(ab^2)^{-3}}{(x^2y^{-3})^4}$

5. $\dfrac{(3ab^5)^{-3}}{2x^{-5}}$

6. $\dfrac{7x^{-1}}{y^2}$

7. $(5w^{-2})^2(2w^{-2})$

8. $\dfrac{x^{-2}y^{-3}}{(c)^{-2}}$

9. $\dfrac{(5a^2b^3)^2}{(-2x)^{-3}}$

10. $\dfrac{16w^{-1}y^2z^{-3}}{2x}$

– –

If additional practice is needed:
  Leithold, page 102, problems 1–10 and 29–34
  Peters, page 87, problems 1–8, 11, 13
  Rees, Sparks, and Rees, page 386, problems 1–16

UNIT 10

Division of Powers

In this unit you will learn to simplify expressions in which variables with exponents appear in both the numerator and the denominator. When you have completed the unit, you will be able to simplify expressions involving **division of like variables** raised to different positive integral exponents.

Recall the four laws of exponents from Unit 8.

Laws of Exponents		
I.	**Multiplication**	$b^n \cdot b^m = b^{n+m}$
II.	**Power of a power**	$(b^n)^m = b^{nm}$
III.	**Power of a product**	$(bc)^n = b^n c^n$
IV.	**Power of a fraction**	$\left(\dfrac{a}{b}\right)^n = \dfrac{a^n}{b^n}$

We will now add a fifth and final law of exponents, which deals with division:

$$\text{V.} \quad \textbf{Division:} \quad \frac{b^m}{b^n} = b^{m-n}$$

$$\text{or, alternatively,}$$

$$= \frac{1}{b^{n-m}}$$

From a brief analysis of the laws we can produce a convenient way to classify them.

I.   Deals with multiplication—exponents added.

II. ⎫
III. ⎬ "In a sense" deal with removing parentheses—exponents multiplied.
IV. ⎭

V.   Deals with division—exponents subtracted.

EXAMPLE 1

Simplify: $\dfrac{x^5}{x^3}$.

Solution: $\dfrac{x^5}{x^3} = x^{5-3} = x^2$

An alternative and longer solution uses the definition of integral exponents:

$$\frac{x^5}{x^3} = \frac{x \cdot x \cdot \cancel{x} \cdot \cancel{x} \cdot \cancel{x}}{\cancel{x} \cdot \cancel{x} \cdot \cancel{x}} = x \cdot x = x^2$$

EXAMPLE 2

Simplify: $\dfrac{a^4}{a^3}$.

Solution: $\dfrac{a^4}{a^3} = a^{4-3} = a$

What about

$$\frac{c^{15}}{c^3} = ?$$

So far we have been using only the first half of the division law of exponents. Now consider a situation where the exponent in the denominator is larger than the exponent in the numerator.

EXAMPLE 3

Simplify: $\dfrac{x}{x^4}$.

Solution: Method 1. $\dfrac{x}{x^4} = x^{1-4} = x^{-3} = \dfrac{1}{x^3}$

Method 2. $\dfrac{x}{x^4} = \dfrac{1}{x^{4-1}} = \dfrac{1}{x^3}$

When the larger exponent is in the denominator, it is faster to use Method 2—subtracting in the denominator.

EXAMPLE 4

Simplify: $\dfrac{a^2}{a^{10}}$.

Solution: $\dfrac{a^2}{a^{10}} = \dfrac{1}{a^{10-2}} = \dfrac{1}{a^8}$

Now consider these examples, in which both the numerator and the denominator are at times larger.

EXAMPLE 5

Simplify: $\dfrac{a^2 b^7 c^3}{a^5 b^2 c^4}$.

Solution: $\dfrac{a^2 \, b^7 \, c^3}{a^5 \, b^2 \, c^4} = \dfrac{b^{7-2}}{a^{5-2} c^{4-3}} = \dfrac{b^5}{a^3 c}$

EXAMPLE 6

Simplify: $\dfrac{a^2 b c^3}{a^7 b^3 c^3}$.

Solution: $\dfrac{a^2 \, b \, c^3}{a^7 \, b^3 \, c^3} = \dfrac{1}{a^{7-2} b^{3-1} c^{3-3}} = \dfrac{1}{a^5 b^2}$

The c does not appear since $c^{3-3} = c^0 = 1$.

Again, let's move on to some problems involving all the laws of exponents.

Recall that, when simplifying an expression, we are attempting to write it without using any zero or negative exponents or parentheses.

To accomplish this, I generally:

Remove the parentheses.
Write with positive exponents.
Use the **division** law, V.
Use the **multiplication** law, I.

There is nothing sacred about the order of these steps. In fact, they are completely interchangeable. You may do them in any order you wish. I suggest only that you *establish a pattern of your own and stick with it.*

EXAMPLE 7

Simplify: $\left(\dfrac{xy^0}{x^{-5}}\right)^{-2}$

- -

Solution: $\dfrac{x^{-2}y^0}{x^{10}} = \dfrac{x^{-2}}{x^{10}} = \dfrac{1}{x^2 x^{10}} = \dfrac{1}{x^{12}}$

- -

EXAMPLE 8

Simplify: $\dfrac{(2^2)^{-1}}{(2^{-4})^2}$.

- -

Solution: $\dfrac{(2^2)^{-1}}{(2^{-4})^2} = \dfrac{2^{-2}}{2^{-8}} = \dfrac{2^8}{2^2} = 2^{8-2} = 2^6$

- -

EXAMPLE 9

Simplify: $\left(\dfrac{x^2 y}{xy^{-4}}\right)^3$

- -

Solution: $\dfrac{x^6 y^3}{x^3 y^{-12}} = \dfrac{x^6 y^3 y^{12}}{x^3} = x^{6-3}y^3 y^{12} = x^3 y^{15}$

- -

Now try two problems yourself.

Problem 1

Simplify: $\left[\dfrac{(3x^2y)^3}{3x^7y^9}\right]^2$.

Solution: Try to simplify the expression without looking at the solution below.

(Hint: Since there are two sets of grouping symbols, work, as always, from the inside out to remove them.)

$$\left[\frac{(3x^2y)^3}{3x^7y^9}\right]^2 = \left[\frac{3^3x^6y^3}{3x^7y^9}\right]^2 = \frac{3^6x^{12}y^6}{3^2x^{14}y^{18}} = \frac{3^{6-2}}{x^{14-12}y^{18-6}} = \frac{3^4}{x^2y^{12}}$$

Answer: $\dfrac{3^4}{x^2y^{12}}$

Problem 2

Simplify: $\left(\dfrac{2x^{-1}y^2}{x^{-3}}\right)^2$.

Solution:

Answer: $4x^4y^4$

You should now be able to simplify expressions in which it is necessary to divide terms with positive and zero exponents. You should now also be able to apply all five laws of exponents.

Laws of Exponents

I.   **Multiplication** $b^n \cdot b^m = b^{n+m}$

II. **Power of a power** $(b^n)^m = b^{nm}$

III. **Power of a product** $(bc)^n = b^n c^n$

IV. **Power of a fraction** $\left(\dfrac{a}{b}\right)^n = \dfrac{a^n}{b^n}$

V.   **Division** $\dfrac{b^m}{b^n} = b^{m-n}$

                               or, alternatively,

                               $= \dfrac{1}{b^{n-m}}$

The classification scheme we developed for these laws will help you retain and use them.

I.   Deals with multiplication—exponents added.

II. ⎫
III. ⎬ "In a sense" deal with removing parentheses—exponents multiplied.
IV. ⎭

V.   Deals with division—exponents subtracted.

Finally, I recommend establishing, and adhering to, your own pattern for simplification of expressions with exponents, such as mine on page 64.

Now try to simplify the following expressions.

EXERCISES

Simplify (write without using negative or zero exponents):

1. $\dfrac{x^{-4}}{x^4}$

2. $\dfrac{15x^5 y^3}{3x^2 y^7}$

3. $\dfrac{x^5 \cdot x^{-4}}{x^{-3}}$

4. $x(3x^2y^{-3})^2$

5. $(2w^{-2})^2(5w^{-2})$

6. $x(5xy^{-2})^{-2}$

7. $\dfrac{7a^{-4}b^4}{28a^{-3}b^{-3}}$

8. $\dfrac{(3x^2y)^{-1}}{2xy^{-5}}$

9. $\dfrac{(m^{-3}s^{-3})^3}{m^{-4}s^4}$

10. $\dfrac{3x^{-2}y^3c^5}{x^{-3}y^7c^2}$

11. $\dfrac{(3xy^{-2})^{-3}}{x}$

12. $\left(\dfrac{a^{-1}b^2}{a^{-3}b^{-2}}\right)^{-2}$

13. $\left[\dfrac{x^{-4}y^{-3}z^2}{x^{-3}y^2z^{-4}}\right]^{-2}$

14. $\left[\dfrac{(ab)^{-1}}{(a^{-2}b^3)^3}\right]^{-1}$ If you can do this one, you've got it made!

––

If additional practice is needed:
Lial and Miller, pages 41 and 42, problems 9–18
Leithold, page 102, problems 20–28, 35–37
Rees, Sparks, and Rees, page 386, problems 17–36

UNIT 11

Review of Fractions: Addition and Subtraction

In the preceding units we discussed positive integral exponents as well as zero and negative exponents. Before learning about **fractional** exponents, we need to review some things about fractions themselves.

In this unit we will review how fractions are added and subtracted. When you have completed the unit, you will be able to solve problems involving the addition and subtraction of fractions, without changing the terms to decimals.

Before we begin, we must be certain of several definitions:

Definition: Let a and b be integers with $b \neq 0$. Then $\frac{a}{b}$ is called a **rational number** (generally referred to as a **fraction**).

   a is the numerator   and b is the denominator.

Definition: A **factor** is a number or letter that is being **multiplied**.

EXAMPLE 1

Consider: $3ax$. Since $3ax$ means $3 \cdot a \cdot x$, there are three factors: 3, a, and x.

Definition: A fraction $\frac{a}{b}$ is said to be in **lowest terms** when all possible common factors **in the numerator and denominator** have been canceled.

EXAMPLE 2

Consider: $\quad \frac{6}{15} = \frac{2 \cdot 3}{5 \cdot 3} = \frac{2}{5}; \quad \frac{2}{5}$ is in lowest terms.

Before proceeding, we need to comment on the **sign** of a fraction.

$$\frac{-a}{b} = \frac{a}{-b} = -\frac{a}{b}$$

These three are equivalent fractions! The negative sign may be placed either in the numerator, in the denominator, or in front of the entire fraction. My preference is to place the negative sign in the numerator, which is what is done throughout this book.

Let me add a note of caution before proceeding. Please do *not* change the factions to decimals, or you will defeat the intention of this and the following unit, which is to enable you to work effectively with algebraic as well as numerical fractions.

ADDITION

Definition: **Addition of Rationals with Common Denominator**

$$\frac{a}{b} + \frac{c}{b} = \frac{a + c}{b}$$

If the fractions have the same denominator, **add the numerators**. The denominator will remain the same.

EXAMPLE 3

Add: $\frac{1}{7} + \frac{2}{7}$.

Solution: $\quad \frac{1}{7} + \frac{2}{7} = \frac{3}{7}$

EXAMPLE 4

Add: $\dfrac{2}{5} + \dfrac{4}{5}$.

Solution: $\dfrac{2}{5} + \dfrac{4}{5} = \dfrac{2 + 4}{5} = \dfrac{6}{5}$

Note: Leave the answer as an improper fraction; it is neither necessary nor advisable to change it to $1\frac{1}{5}$.

Unfortunately, few problems ever occur in which the fractions have the same denominator. However, the following definition allows us to add two fractions in one step without ever bothering to find the lowest common denominator. I rather like this particular definition, and I use it all the time.

Definition: **Addition of Rationals**

$$\dfrac{a}{b} + \dfrac{c}{d} = \dfrac{ad + bc}{bd}$$

Because of this definition, it is *not* necessary to find the least common denominator.

EXAMPLE 5

Add: $\dfrac{7}{8} + \dfrac{2}{3}$.

Solution: $\dfrac{7}{8} + \dfrac{2}{3} = \dfrac{7 \cdot 3 + 8 \cdot 2}{8 \cdot 3} = \dfrac{21 + 16}{24} = \dfrac{37}{24}$

EXAMPLE 6

Add: $\dfrac{2}{3} + \dfrac{1}{5}$.

Solution: $\dfrac{2}{3} + \dfrac{1}{5} = \dfrac{10 + 3}{15} = \dfrac{13}{15}$

Now try two problems yourself.

Problem 1

Add: $\dfrac{4}{7} + \dfrac{5}{6}$.

Solution:

Answer: $\dfrac{59}{42}$

Problem 2

Add: $\dfrac{-2}{3} + \dfrac{2}{5}$.

Solution:

Answer: $-\dfrac{4}{15}$

The advantage of the definition given above is that for simple fractions the addition can be performed mentally. However, it can be used only when adding two fractions; if there are more, the fractions must be added two at a time with this definition.

SUBTRACTION

Definition: **Subtraction of Rationals**
$\dfrac{a}{b} - \dfrac{c}{d} = \dfrac{ad - bc}{bd}$

Note that the only difference between this definition and the definition for the addition of rationals is the minus sign.

But be careful—the first term of the numerator must be the product found by multiplying diagonally to the right and **down**.

$$\frac{a}{b} \searrow \frac{c}{d} = \frac{ad - }{}$$

EXAMPLE 7

Subtract: $\dfrac{3}{5} - \dfrac{2}{3}$.

Solution: $\dfrac{3}{5} \searrow \dfrac{2}{3} = \dfrac{9 - 10}{15} = \dfrac{-1}{15}$

EXAMPLE 8

Subtract: $\dfrac{-5}{11} - \dfrac{3}{4}$.

Solution: $\dfrac{-5}{11} \searrow \dfrac{3}{4} = \dfrac{-20 - 33}{44} = \dfrac{-53}{44}$

EXAMPLE 9

Subtract: $\dfrac{1}{3} - \dfrac{1}{2}$.

Solution: $\dfrac{1}{3} \searrow \dfrac{1}{2} = \dfrac{2 - 3}{6} = \dfrac{-1}{6}$

EXAMPLE 10

Solve: $\dfrac{1}{6} + \dfrac{-2}{3}$.

Solution: $\dfrac{1}{6} \searrow \dfrac{-2}{3} = \dfrac{3 - 12}{18} = \dfrac{-9}{18} = \dfrac{-1}{2}$

Note: After using the definition, our solution was not in the lowest terms and had to be reduced in the final step.

Try a subtraction problem.

Problem 3

Solve: $\dfrac{-3}{4} - \dfrac{2}{5}$.

- -

Solution:

Answer: $\dfrac{-23}{20}$

- -

With algebraic fractions the entire process becomes even easier.

EXAMPLE 11

Add: $\dfrac{1}{x} + \dfrac{1}{y}$.

- -

Solution: $\dfrac{1}{x} + \dfrac{1}{y} = \dfrac{y + x}{xy}$

- -

EXAMPLE 12

Add: $\dfrac{2x}{y} - \dfrac{z}{3}$.

- -

Solution: $\dfrac{2x}{y} \quad \dfrac{z}{3} = \dfrac{6x - yz}{3y}$

- -

You should now be able to add and subtract rationals (fractions). Before beginning the next unit, you should solve the following problems involving rationals. Reduce the answers to the lowest terms, but *do not change them to decimals*.

EXERCISES

Solve:

1. $\dfrac{2}{11} + \dfrac{1}{11}$

2. $\dfrac{7}{10} - \dfrac{9}{10}$

3. $\dfrac{7}{9} + \dfrac{1}{5}$

4. $\dfrac{2}{13} + \dfrac{6}{-5}$

5. $\dfrac{-1}{2} + \dfrac{1}{4}$

6. $\dfrac{-6}{-15} + \dfrac{3}{5}$

7. $\dfrac{2}{9} - \dfrac{-1}{10}$

8. $\dfrac{11}{t} + \dfrac{7}{r}$

9. $\dfrac{1}{x} + \dfrac{1}{x}$

10. $\dfrac{10}{x+1} + \dfrac{3}{x+1}$

11. $\dfrac{5}{a} - \dfrac{4}{a}$

12. $\dfrac{-s}{9} + \dfrac{k}{10}$

13. $\dfrac{x}{2} - \dfrac{x}{5}$

14. $\dfrac{x+1}{2} - \dfrac{3}{5}$

15. $\dfrac{x-1}{3} - \dfrac{x+1}{2}$

16. $\dfrac{x+2}{2} - \dfrac{x+3}{3}$

If additional practice is needed:
Peters, pages 69 and 70, problems 1–12
Rich, pages 198 and 199, problems 20, 21, 24

UNIT 12

Review of Fractions: Multiplication and Division; Complex Fractions

Here, as in Unit 11, we will review some facts about **fractions** before we study **fractional exponents**. When you have completed this unit, you will be able to solve problems involving the multiplication and division of fractions, without changing the terms to decimals.

MULTIPLICATION

Definition: **Multiplication of Rationals**

$$\frac{a}{c} \cdot \frac{c}{d} = \frac{ac}{bd}$$

Multiply numerators together, and multiply denominators together.

Note: 1. You do not need a common denominator.
2. You should factor and cancel as soon as possible.

EXAMPLE 1

Multiply: $\dfrac{2}{3} \cdot \dfrac{4}{5}$.

Solution: $\dfrac{2}{3} \cdot \dfrac{4}{5} = \dfrac{2 \cdot 4}{3 \cdot 5} = \dfrac{8}{15}$

EXAMPLE 2

Multiply: $\dfrac{4}{7} \cdot \dfrac{35}{12}$.

Solution: $\dfrac{\overset{1}{\cancel{4}}}{7} \cdot \dfrac{\overset{5}{\cancel{35}}}{\underset{3}{\cancel{12}}} = \dfrac{5}{3}$

EXAMPLE 3

Multiply: $\dfrac{12}{15} \cdot \dfrac{5}{18}$.

Solution: $\dfrac{\overset{2}{\cancel{12}}}{\underset{3}{\cancel{15}}} \cdot \dfrac{\overset{1}{\cancel{5}}}{\underset{3}{\cancel{18}}} = \dfrac{2 \cdot 1}{3 \cdot 3} = \dfrac{2}{9}$

DIVISION

Definition: **Division of Rationals**
$$\frac{a}{b} \div \frac{e}{f} = \frac{a}{b} \cdot \frac{f}{e} = \frac{af}{be}$$

When dividing by a fraction, we invert the second fraction (the divisor) and multiply.

EXAMPLE 4

Divide: $\dfrac{2}{3} \div \dfrac{-3}{8}$.

Solution: $\dfrac{2}{3} \div \dfrac{-3}{8} = \dfrac{2}{3} \cdot \dfrac{8}{-3} = \dfrac{16}{-9} = \dfrac{-16}{9}$

Here are three problems for you.

Problem 1

Simplify: $\dfrac{2}{-5} \cdot \dfrac{30}{8}$.

Solution:

Answer: $-\dfrac{3}{2}$

Problem 2

Simplify: $\dfrac{2}{-5} \div \dfrac{30}{8}$.

Solution:

Answer: $-\dfrac{8}{75}$

Problem 3

Simplify: $\dfrac{-12}{20} \div \dfrac{10}{-16}$.

Solution:

Answer: $\dfrac{24}{25}$

COMPLEX FRACTIONS

We must now consider **complex fractions**—fractions in which there are one or more fractions in the numerator or denominator, or in both. Remember that in a complex fraction, as in a simple fraction, the horizontal bar means simply that we should divide. For example:

$$\frac{12}{2} \text{ means 12 divided by 2}$$

EXAMPLE 5

Simplify this complex fraction: $\dfrac{\frac{2}{5}}{3}$.

Solution: $\dfrac{\frac{2}{5}}{3} = \dfrac{2}{5} \div 3 \quad$ Rewrite, using a division sign.

$$= \frac{2}{5} \cdot \frac{1}{3}$$

$$= \frac{2}{15}$$

EXAMPLE 6

Simplify: $\dfrac{\frac{1}{2} + \frac{2}{3}}{\frac{5}{4}}$.

Solution: $\dfrac{\frac{1}{2} + \frac{2}{3}}{\frac{5}{4}} = \left(\dfrac{1}{2} + \dfrac{2}{3}\right) \div \dfrac{5}{4}$ Rewrite, using a division sign.

$\qquad = \left(\dfrac{3 + 4}{6}\right) \div \dfrac{5}{4}$ Add fractions in parentheses.

$\qquad = \dfrac{7}{6} \cdot \dfrac{4}{5}$

$\qquad = \dfrac{7}{\overset{}{\underset{3}{6}}} \cdot \dfrac{\overset{2}{4}}{5}$

$\qquad = \dfrac{14}{15}$

EXAMPLE 7

Simplify: $\dfrac{3 + \frac{1}{x}}{2}$.

Solution: $\dfrac{3 + \frac{1}{x}}{2} = \left(3 + \dfrac{1}{x}\right) \div 2$ Rewrite, using a division sign.

$\qquad = \left(\dfrac{3}{1} + \dfrac{1}{x}\right) \div \dfrac{2}{1}$ $3 = \dfrac{3}{1}$.

$\qquad = \dfrac{3x + 1}{x} \cdot \dfrac{1}{2}$ Add fractions in parentheses.

$\qquad = \dfrac{3x + 1}{2x}$

Can you do this one?

Problem 4

Simplify: $\dfrac{x + \frac{1}{2}}{\frac{1}{2}}$.

- -

Solution:

Answer: $2x + 1$

- -

Here are three more examples to end the unit.

===

EXAMPLE 8

Simplify: $\dfrac{5 - \dfrac{3}{x}}{x}$.

- -

Solution: $\dfrac{5 - \dfrac{3}{x}}{x} = \left(\dfrac{5}{1} - \dfrac{3}{x}\right) \div x$

$$= \left(\dfrac{5x - 3}{x}\right) \cdot \dfrac{1}{x}$$

$$= \dfrac{(5x - 3) \cdot 1}{x \cdot x}$$

$$= \dfrac{5x - 3}{x^2}$$

- -

EXAMPLE 9

Simplify: $\dfrac{5}{\dfrac{1}{x}+2}$.

Solution: $\dfrac{5}{\dfrac{1}{x}+2} = 5 \div \left(\dfrac{1}{x} + 2 \right)$

$$= \dfrac{5}{1} \div \left(\dfrac{1}{x} + \dfrac{2}{1} \right)$$

$$= \dfrac{5}{1} \div \left(\dfrac{1+2x}{x} \right)$$

$$= \dfrac{5}{1} \cdot \dfrac{x}{1+2x}$$

$$= \dfrac{5 \cdot x}{1 \cdot (1+2x)}$$

$$= \dfrac{5x}{1+2x}$$

EXAMPLE 10

Simplify: $\dfrac{\dfrac{x+1}{x} - \dfrac{1}{3}}{\dfrac{1}{x}}$.

Solution: $\dfrac{\dfrac{x+1}{x} - \dfrac{1}{3}}{\dfrac{1}{x}} = \left(\dfrac{x+1}{x} - \dfrac{1}{3} \right) \div \dfrac{1}{x}$

$$= \dfrac{3(x+1) - x}{3x} \cdot \dfrac{x}{1}$$

$$= \dfrac{3x+3-x}{3x} \cdot \dfrac{x}{1}$$

$$= \dfrac{2x+3}{3\cancel{x}} \cdot \dfrac{\cancel{x}}{1}$$

$$= \dfrac{2x+3}{3}$$

You should now be able to multiply and divide fractions. You should also be able to simplify complex fractions in which there are fractions in the numerator and/or denominator.

Before beginning the next unit, do the following exercises, reducing the answers to the lowest terms without *converting to decimals*.

EXERCISES

Simplify:

1. $\dfrac{4}{7} \cdot \dfrac{35}{12}$

2. $\dfrac{5}{18} \div \dfrac{3}{14}$

3. $\dfrac{3}{4} \cdot \dfrac{-8}{9}$

4. $\dfrac{9}{14} \div \dfrac{5}{21}$

5. $\dfrac{-8}{9} \div \dfrac{12}{-7}$

6. $\left(\dfrac{2}{3} \cdot \dfrac{4}{5} \right) \cdot \dfrac{10}{10}$

7. $\dfrac{\frac{3}{10}}{\frac{1}{10}}$

8. $\dfrac{3 - \frac{2}{5}}{3 + \frac{2}{5}}$

9. $\dfrac{1 - \frac{1}{3}}{\frac{5}{6}}$

10. $\dfrac{\frac{a}{2} - \frac{3}{5}}{2}$

11. $\dfrac{\frac{7}{x-1}}{4}$

12. $\dfrac{\frac{2}{x} - 5}{x}$

13. $\dfrac{9 - \dfrac{x}{4}}{\dfrac{1}{2}}$

14. $\dfrac{\dfrac{x}{5} - \dfrac{x+2}{2}}{\dfrac{x}{2}}$

15. $\dfrac{\dfrac{a}{b} + 2}{\dfrac{a}{b} + 1}$

16. $\dfrac{\dfrac{4}{\dfrac{3}{2} - \dfrac{x+1}{x}}}{}$

_ _

If additional practice is needed:
  Peters, page 76, problems 1–7, 10, 13, 15
  Rich, pages 196–200, problems 15, 18, 29

UNIT 13

Fractional Exponents

In Units 11 and 12 we reviewed how to add, subtract, multiply, and divide fractions. We are now ready to proceed to handling the most difficult exponents, **fractional exponents**.

First recall how we define other exponents:

Positive Integer: $b^n = \underbrace{b \cdot b \cdot b \cdot \ldots \cdot b}_{n \text{ factors}}$

Zero: $b^0 = 1$

Negative: $b^{-n} = \dfrac{1}{b^n}$ and $\dfrac{1}{b^{-n}} = b^n$

Now we can define a fractional exponent.

> Definition: **Fractional Exponent**
> $$b^{n/d} = \left(\sqrt[d]{b}\right)^n = \sqrt[d]{b^n}$$

Note: $b^{n/d}$ ⟵ denotes power

$b^{n/d}$ ⟵ denotes the root

Note that two forms of the definition are given.

$$b^{n/d} = (\sqrt[d]{b})^n = \sqrt[d]{b^n}$$

$$\uparrow \qquad \uparrow$$
$$\text{numerical} \quad \text{algebraic}$$

The first form is useful in numerical calculations, provided that the dth root of b is a known integer. It is then convenient to take the root before raising to the nth power in order to work with smaller numbers.

The second form is the more common way of rewriting algebraic expressions with fractional exponents.

As the two forms suggest, we can do either the root or the power first. Here we will consider some numerical problems first.

Again, consider the definition in the form for numerical problems.

Fractional Exponent with Numerical Problems

$$\text{power} \quad \text{root}$$
$$\downarrow \quad \downarrow$$
$$b^{n/d} = (\sqrt[d]{b})^n$$
$$\uparrow$$
$$\text{numerical}$$

EXAMPLE 1 $8^{1/3} = (\sqrt[3]{8})^1 = (2)^1 = 2$

EXAMPLE 2 $4^{1/2} = (\sqrt{4})^1 = (2)^1 = 2$

Note that the 2 is not usually written for square roots.

EXAMPLE 3 $4^{3/2} = (\sqrt{4})^3 = (2)^3 = 8$

EXAMPLE 4 $27^{2/3} = (\sqrt[3]{27})^2 = (3)^2 = 9$

EXAMPLE 5 $8^{4/3} = (\sqrt[3]{8})^4 = (2)^4 = 16$

EXAMPLE 6 $9^{-1/2} = (\sqrt{9})^{-1} = (3)^{-1} = \dfrac{1}{3}$

or, alternatively,

$$= \frac{1}{9^{1/2}} = \frac{1}{\sqrt{9}} = \frac{1}{3}$$

Now you try to simplify some numerical expressions with fractional exponents.

Problem 1 $32^{1/5} =$

Problem 2 $(-1)^{1/3} =$

Problem 3 $4^{5/2} =$

Problem 4 $64^{2/3} =$

Problem 5 $81^{3/4} =$

Problem 6 $(-8)^{2/3} =$

Answers: 1. 2   2. –1   3. 32   4. 16   5. 27   6. 4

And now for some problems where it is not possible to determine the root. As you may suspect, these cannot be simplified much without using tables or calculators.

EXAMPLE 7 $5^{2/3} = \sqrt[3]{5^2} = \sqrt[3]{25}$

EXAMPLE 8 $7^{1/5} = \sqrt[5]{7}$

Now consider the definition of a fractional exponent in the form for algebraic expressions:

Fractional Exponent with Algebraic Expressions

$$\overset{\text{power root}}{\overset{\downarrow \ \downarrow}{b^{n/d}}} = \sqrt[d]{b}^{\,n}$$

algebraic

EXAMPLE 9 $x^{2/3} = \sqrt[3]{x^2}$

EXAMPLE 10 $y^{2/5} = \sqrt[5]{y^2}$

The only time any simplifying can be done is if n/d is an improper fraction. The following examples illustrate the procedure then to be used.

EXAMPLE 11 $x^{3/2} = \sqrt{x^3} = \sqrt{x^2 \cdot x} = x\sqrt{x}$

The final answer is preferred to $\sqrt{x^3}$.

EXAMPLE 12 $x^{7/5} = \sqrt[5]{x^7} = \sqrt[5]{x^5 \cdot x^2} = x\sqrt[5]{x^2}$

EXAMPLE 13 $x^{4/3} = \sqrt[3]{x^4} = \sqrt[3]{x^3 \cdot x} = x\sqrt[3]{x}$

EXAMPLE 14 $x^{5/2} = \sqrt{x^5} = \sqrt{x^4 \cdot x} = x^2\sqrt{x}$

The following numerical problems should be more familiar to you.

EXAMPLE 15 $\sqrt{12} = \sqrt{4 \cdot 3} = 2\sqrt{3}$

EXAMPLE 16 $\sqrt{50} = \sqrt{25 \cdot 2} = 5\sqrt{2}$

You should now be able to simplify expressions involving fractional exponents, whether they occur in numerical or in algebraic problems. Note that we are concerned mainly with the ability to handle the fractional exponents and are *not* concerned about radicals here.

Before beginning the next unit, you should simplify the following expressions involving fractional exponents.

EXERCISES

Simplify:

1. $32^{1/5}$

2. $(-1)^{2/3}$

3. $(-4)^{1/2}$ Careful!

4. $4^{3/2}$

5. $4^{1/2}$

6. $4^{-1/2}$

7. $x^{-1/2}$

8. $x^{1/3}$

9. $a^{2/5}$

10. $4^{-3/2}$

11. $(x+1)^{1/2}$

12. $x^{8/3}$

13. $(4x)^{1/2}$

14. $x^{11/2}$

15. $(5x)^{-1/2}$

16. $\sqrt{18x^3}$

17. $(2x)^{2/3}$

18. $(-64)^{2/3}$

19. Rewrite using a fractional exponent: $\sqrt{7x}$.

20. Rewrite using a fractional exponent: $\sqrt[3]{2x}$.

UNIT 14

Simplifying Expressions with Fractional Exponents

You have completed all the definitions and laws of exponents necessary to handle any given expression involving exponents. From here on, you need practice, confidence, and a bit of luck! We're going to work on practice in this unit.

The definitions and laws of exponents are restated below for reference:

Positive Integer: $b^n = \underbrace{b \cdot b \cdot b \cdot \;\cdots\; \cdot b}_{n \text{ factors}}$

Zero: $b^0 = 1$

Negative: $b^{-n} = \dfrac{1}{b^n}$ and $\dfrac{1}{b^{-n}} = b^n$

Fractional: $b^{n/d} = \sqrt[d]{b^n} = (\sqrt[d]{b})^n$

Laws of Exponents

I. **Multiplication** $b^n \cdot b^m = b^{n+m}$

II. **Power of a power** $(b^n)^m = b^{nm}$

III. **Power of a product** $(bc)^n = b^n c^n$

IV. **Power of a fraction** $\left(\dfrac{a}{b}\right)^n = \dfrac{a^n}{b^n}$

V. **Division** $\dfrac{b^m}{b^n} = b^{m-n}$

$= \dfrac{1}{b^{n-m}}$

Luckily, fractional exponents obey all the laws of exponents.

To simplify expressions involving exponents means to write the final answer without using zero, negative, or fractional exponents.

Here we get into some dandy problems. To make things a bit clearer, I have tried to follow a basic pattern when working all problems:

Remove parentheses.
Write with positive exponents.
Use the **division** law of exponents.
Finish with the **multiplication** law of exponents.

It is the same basic pattern I have been using since Unit 10.

EXAMPLE 1	$(x^{-1/2})^{-2/3} = x^{1/3} = \sqrt[3]{x}$	Law II
EXAMPLE 2	$x^{-1/3} \cdot x^{1/2} = x^{(-1/3)+1/2} = x^{(-2+3)/6} = x^{1/6} = \sqrt[6]{x}$	Law I
EXAMPLE 3	$(9x^{-4})^{1/2} = 9^{1/2}x^{-2} = \dfrac{3}{x^2}$	Laws III and II
EXAMPLE 4	$(x^6y^{-3})^{-2/3} = x^{-4}y^2 = \dfrac{y^2}{x^4}$	Laws II and III

You try one; then we'll have some more examples.

Problem 1

Simplify: $(x^{-8}y^4)^{-3/2}$.

Solution:

Answer: $\dfrac{x^{12}}{y^6}$

EXAMPLE 5

Simplify: $\left(\dfrac{a^{1/2}b^{2/3}}{c^{1/7}}\right)^6$.

Solution: $\left(\dfrac{a^{1/2}b^{2/3}}{c^{1/7}}\right)^6 = \dfrac{a^3b^4}{c^{6/7}} = \dfrac{a^3b^4}{\sqrt[7]{c^6}}$

At this point I'm not at all concerned about rationalizing that denominator. Recall from high school days that it was a "no-no" to leave the radical in the denominator.

Quite often expressions you are trying to simplify contain radicals. Rather than trying to learn a complete set of new definitions and laws, it is possible to use the definition of a fractional exponent to rewrite the expression using exponents, and then proceed to simplify.

EXAMPLE 6

Simplify: $\sqrt{x} \cdot \sqrt[3]{x}$.

Solution: $\sqrt{x} \cdot \sqrt[3]{x} = x^{1/2} \cdot x^{1/3} = x^{1/2 + 1/3} = x^{5/6} = \sqrt[6]{x^5}$

EXAMPLE 7

Simplify: $\dfrac{\sqrt[3]{x^2}}{\sqrt[4]{x}}$.

Solution: $\dfrac{x^{2/3}}{x^{1/4}} = x^{2/3 - 1/4} = x^{(8-3)/12} = x^{5/12} = \sqrt[12]{x^5}$

EXAMPLE 8

Simplify: $\left[\left(\dfrac{\sqrt{2}}{-3}\right)^{-4}\right]^{-1}$.

Solution: $\left[\left(\dfrac{(2)^{1/2}}{-3}\right)^{-4}\right]^{-1} = \left[\dfrac{(2)^{-2}}{(-3)^{-4}}\right]^{-1} = \dfrac{2^2}{(-3)^4} = \dfrac{4}{81}$

As always, work from the inside out to remove parentheses.

Here are four problems for you.

Problem 2

Simplify: $(\sqrt[3]{7})^6$.

Solution:

Answer: 49

Problem 3

Simplify: $(x^{1/3}\sqrt{x})^3$.

Solution:

Answer: $x^2\sqrt{x}$

Problem 4

Simplify: $\left(\dfrac{2^0}{8^{1/3}}\right)^{-1}$.

Solution:

Answer: 2

Problem 5

Simplify: $(2x\sqrt[6]{x})^3$.

Solution:

Answer: $8x^3\sqrt{x}$

In case you had difficulty with Problems 2–5, let me restate our objective and my procedure.

Objective: To be able to simplify expressions involving exponents. By "simplify" we mean to write the expression without using parentheses or fractional, negative, or zero exponents.

My procedure (There is nothing sacred about the order of the steps. I am merely listing them so that you can follow what I am doing in the solutions.):

1.   If there are radicals, rewrite using fractional exponents.
2.   Remove parentheses (Law II, III, or IV).
3.   Write with positive exponents.
4.   Use the division or multiplication law.
5.   If any fractional exponents remain, rewrite using radicals.

The following example illustrates this procedure one more time.

EXAMPLE 9

Simplify: $\left(\dfrac{\sqrt{x}}{x^2}\right)^{-2}$

Solution: $\left(\dfrac{\sqrt{x}}{x^2}\right)^{-2}$

Step 1. $\left(\dfrac{x^{1/2}}{x^2}\right)^{-2}$

Step 2. $\dfrac{x^{-1}}{x^{-4}}$

Step 3. $\dfrac{x^4}{x}$

Step 4. x^3

You should now be able to simplify most expressions involving exponents, whether they are fractional, negative, zero, or integers, as well as being able to simplify expressions with radicals.

Before beginning the next unit you should simplify the expressions in the exercises. The final answers should be written without fractional, negative, or zero exponents. Also, fractions should be in lowest terms. When you have completed the exercises, check your answers against those given at the back of the book. Because these problems are difficult, the complete solutions are provided for all fifteen.

EXERCISES

Simplify:

1. $\dfrac{y^{2/3}}{y^{1/3}}$

2. $(y^{3/5})^{1/4}$

3. $x^{1/2} \cdot x^{3/5}$

4. $\left(\dfrac{a^4}{c^2}\right)^{1/2}$

5. $[(3\sqrt{4})^{-1}]^2$

6. $(x\sqrt{x})^{1/2}$

7. $(8x^2)^{1/3}$

8. $\left(\dfrac{2^{-3} \cdot 2^5}{2^{-2}}\right)^3$

9. $\left(\dfrac{x^{1/3}}{x^{2/3}}\right)^3$

10. $x^{1/2} \cdot x^{5/2}$

11. $(\sqrt[3]{x^2})^{1/2}$

12. $$\dfrac{x^{-7/2} \cdot x^{3/2}}{\sqrt{x} \cdot x^{-3/2}}$$

13. $(8\sqrt{x})^{-2/3}$

14. $\left(\dfrac{27^{5/3} \cdot 27^{-1/3}}{27^{1/3}} \right)^{2}$

15. $\left(\dfrac{3x^{-1}}{\sqrt{x}} \right)^{2}$

If additional practice is needed:
   Leithold, pages 115 and 116, problems 17–36
   Lial and Miller, page 45, problems 11–16, 19–24
   Peters, pages 96–98, problems 1–66, 80–99
   Rees, Sparks, and Rees, page 389, problems 25–44

UNIT 15

Additional Practice with Exponents

Before we leave our study of exponents, we will look at a source of common errors in working with them. These are errors in dealing with the addition and subtraction of terms with exponents.

The five laws of exponents deal only with simplifying multiplication and division. In fact, addition and subtraction involving terms with exponents must be simplified using the definitions covered in Units 11 and 12 on fractions.

EXAMPLE 1

Simplify: $(ab)^{-1}$

- -

Solution $(ab)^{-1} = a^{-1}b^{-1} = \dfrac{1}{ab}$

- -

EXAMPLE 2

Simplify: $(a + b)^{-1}$.

Solution: $(a + b)^{-1} = \dfrac{1}{(a + b)}$

EXAMPLE 3

Simplify: $a^{-1} + b^{-1}$.

Solution: $a^{-1} + b^{-1} = \dfrac{1}{a} \searrow \dfrac{1}{b} = \dfrac{b + a}{ab}$

The arrow is a reminder to add by using the definition from Unit 11.

Be sure you see the difference between Example 1, in which the exponent applies to the product of a and b; Example 2, in which the exponent applies to the sum of a and b; and Example 3, in which the exponents apply to the two separate terms.

EXAMPLE 4

Simplify: $a + b^{-1}$.

Solution: $a + b^{-1} = a + \dfrac{1}{b} = \dfrac{a}{1} \searrow \dfrac{1}{b} = \dfrac{ab + 1}{b}$

Try this problem.

Problem 1

Simplify: $3^{-1} + x^{-1}$.

Solution:

Answer: $\dfrac{x + 3}{3x}$

By now a problem like the following should be easy for you.

Problem 2

Simplify: $\dfrac{a^{-1}b^{-1}}{c^{-1}d^{-1}}$.

- -

Solution:

Answer: $\dfrac{cd}{ab}$

- -

Here's another example.

EXAMPLE 5

Simplify: $\dfrac{a^{-1} + b^{-1}}{c^{-1} + d^{-1}}$.

Solution:
$$\dfrac{\dfrac{1}{a} + \dfrac{1}{b}}{\dfrac{1}{c} + \dfrac{1}{d}} = \left(\dfrac{1}{a} \ \ \dfrac{1}{b}\right) \div \left(\dfrac{1}{c} \ \ \dfrac{1}{d}\right)$$

$$= \dfrac{b + a}{ab} \div \dfrac{d + c}{cd}$$

$$= \dfrac{a + b}{ab} \cdot \dfrac{cd}{c + d}$$

$$= \dfrac{cd(a + b)}{ab(c + d)}$$

We're going to reverse the procedure this time. I'll let you try an easier problem.

Problem 3

Simplify: $\dfrac{1}{a^{-1} + b^{-1}}$.

- -

Solution:

Answer: $\dfrac{ab}{a+b}$

Before proceeding, be sure you understand how Problem 2, Example 5, and Problem 3 differ. Here's a brief explanation:

Problem 2	One term in the expression, factors moved to top or bottom.
Example 5	One fraction, but the numerator has two terms separated by the plus sign and the denominator has two terms also (actually a complex fraction); refer to Unit 12.
Problem 3	Two fractions to be added in the denominator of a complex fraction.

Now consider Examples 6 and 7.

EXAMPLE 6

Simplify: $\dfrac{1}{x^{-1} + x}$.

Solution: $\dfrac{1}{x^{-1}+x} = \dfrac{1}{\dfrac{1}{x}+x}$, which is a complex fraction

$$= 1 \div \left(\frac{1}{x} + x \right)$$

$$= 1 \div \left(\frac{1}{x} + \frac{x}{1} \right)$$

$$= 1 \div \frac{1 + x^2}{x}$$

$$= 1 \cdot \frac{x}{1 + x^2}$$

$$= \frac{x}{1 + x^2}$$

Let me repeat: the laws of exponents deal only with multiplication and division. There are no laws dealing with powers of sums and differences!

EXAMPLE 7

Simplify: $(a + b)^{-2}$.

Solution: $(a + b)^{-2} = \dfrac{1}{(a + b)^2}$ and that is as much as you can do!

You should now be able to simplify expressions in which terms with exponents are added and subtracted. Remember that, since the laws of exponents deal only with multiplication and division, you must call upon the definitions for addition and subtraction of fractions and the techniques for simplifying complex fractions.

Before beginning the next unit, simplify the expressions in the exercises, writing them as either monomials or fractions in lowest terms, with positive exponents only. None of the denominators is zero.

EXERCISES

Simplify:

1. $(x^2 + 1)^{-2}$

2. $x^{-1} + 2^{-2}$

3. $(x^{-1} - 1)^{-1}$

4. $\dfrac{1}{x^{-1} + y^{-1}}$

5. $3^{-1} + 3^{-2}$

6. $\dfrac{x^{-1} + 2y^{-1}}{3}$

7. $\dfrac{a + b^{-1}}{a}$

8. $5(x + y)^{-1}$

9. $\dfrac{x^{-1}}{(2x - 3)^{-2}}$

10. $3x^{-2} + y$

11. $\dfrac{a - a^{-1}}{a + a^{-1}}$

12. $\dfrac{3^{-1} + 2^{-1}}{3^{-1} - 2^{-1}}$

13. $\dfrac{a + b^{-1}}{(ab)^{-1}}$

14. $\dfrac{3ab^{-1}}{a^{-1} + b}$

_____

If additional practice is needed:
   Leithold, page 102, problems 38–45
   Peters, page 87, problems 9–24
   Rees, Sparks, and Rees, page 386, problems 37–42

UNIT 16

Multiplication of Monomials and Polynomials

In this unit we will learn how to multiply binomials and other polynomials. We will discuss some methods that should help you perform these operations and avoid common errors.

Recall that expressions with one term are called **monomials** and that expressions with more than one term (symbols or groups of symbols separated by a plus or minus sign) are called **polynomials.** Polynomials are sometimes classified according to the number of terms; for example, a **binomial** has two terms and a **trinomial** has three terms.

MULTIPLICATION OF TWO MONOMIALS

We will first consider multiplication by a monomial.

Rule: **To multiply two monomials, multiply their numerical coefficients and find the product of the variable factors according to the laws of exponents.**

EXAMPLE 1 $(2x^3)(3x^5) = 2 \cdot 3 \cdot x^3 \cdot x^5 = 6x^8$

EXAMPLE 2 $(2xy^3)(7x^4y^5) = 2 \cdot 7 \cdot x \cdot x^4 \cdot y^3 \cdot y^5 = 14x^5y^8$

EXAMPLE 3 $(3\pi ab)(2\pi a^2bc^3) = 3 \cdot 2 \cdot \pi \cdot \pi \cdot a \cdot a^2 \cdot b \cdot b \cdot c^3 = 6\pi^2a^3b^2c^3$

The above rule extends in the natural way to multiplication of a polynomial by a monomial.

EXAMPLE 4 $2x^3(3x^5 + 2xy^3) = 6x^8 + 4x^4y^3$

EXAMPLE 5 $2xy^3(3x^5 + 7x^4y + 1) = 6x^6y^3 + 14x^5y^4 + 2xy^3$

EXAMPLE 6 $-3a^2b(2a^4b^2 - 6a^3b) = -6a^6b^3 + 18a^5b^2$

Here are three problems for you.

Problem 1 $3xy^2(2xy + 3x^2y^3) =$

Problem 2 $2xyz(x + y + 2z + 3) =$

Problem 3 $-3\pi x\left(2 - 4\pi x^2 + \pi^2 x - \dfrac{\pi}{3}\right) =$

Answers: 1. $6x^2y^3 + 9x^3y^5$ 2. $2x^2yz + 2xy^2z + 4xyz^2 + 6xyz$
3. $-6\pi x + 12\pi^2 x^3 - 3\pi^3 x^2 + \pi^2 x$

MULTIPLICATION OF TWO POLYNOMIALS

Now let's consider multiplication of two polynomials. This is the most frequently encountered problem, and ideally you should be able to do these multiplications mentally.

Consider, for example, $(x + 3)(x + 4)$.

Basically, three methods are commonly used to find the product of two binomials, $(x + 3)(x + 4)$.

Method 1

$$
\begin{array}{r}
x + 3 \\
x + 4 \\
\hline
x^2 + 3x \\
4x + 12 \\
\hline
x^2 + 7x + 12
\end{array}
$$

This way of multiplying polynomials is not very efficient. I generally discourage its use, primarily because it necessitates rewriting the problem.

Method 2 (FOIL)

FOIL for F = first $= x \cdot x$

O = outer $= 4 \cdot x$

I = inner $= 3 \cdot x$

L = last $= 3 \cdot 4$

F   L   F   O   I   L

$(x + 3)(x + 4)$ $=$ $x^2 + 4x + 3x + 12 = x^2 + 7x + 12$

I

O

Another version of the FOIL method looks like this:

x^2 -10 F   O I   L

$(x - 2)(x + 5)$ $=$ $x^2 + 3x - 10$

$-2x$

$5x$

$+3x$

The advantage of the FOIL method is that for simple binomials the multiplication can be performed mentally. However, it can be used only when multiplying two binomials together, *not* when multiplying a binomial by some other polynomial.

Method 3 (Distributive Law)

$(x + 3)(x + 4) = x(x + 4) + 3(x + 4)$

$= x^2 + 4x + 3x + 12$

$= x^2 + 7x + 12$

This method is similar to removing parentheses and works for multiplying *any* type of polynomial. I think it is the most useful method.

EXAMPLE 7

Multiply: $(x - 7)(x + 2)$.

- -

Solution: $(x - 7)(x + 2) = x(x + 2) - 7(x + 2)$

$= x^2 + 2x - 7x - 14$

$= x^2 - 5x - 14$

- -

You try one.

Problem 4

Multiply: $(2x + 5)(3x - 2)$, using either Method 2 or Method 3.

Solution:

Answer: $6x^2 + 11x - 10$

After this example, it will be your turn again.

EXAMPLE 8

Multiply: $(x + a)(2x - b)$.

Solution: $(x + a)(2x - b) = x(2x - b) + a(2x - b)$
$$= 2x^2 - bx + 2ax - ab$$

Problem 5

Multiply: $(3x + 1)(x - 5)$.

Solution:

Answer: $3x^2 - 14x - 5$

Problem 6

Multiply: $(3x - 5)(2a - \pi)$.

Solution:

Answer: $6ax - 3\pi x - 10a + 5\pi$

The next example illustrates the fact that by using Method 3 we can multiply any polynomial by a binomial.

EXAMPLE 9

Multiply: $(x + 4)(x^3 - x^2 + 3x - 1)$.

Solution: $(x + 4)(x^3 - x^2 + 3x - 1) = x(x^3 - x^2 + 3x - 1) + 4(x^3 - x^2 + 3x - 1)$

$$= x^4 - x^3 + 3x^2 - x + 4x^3 - 4x^2 + 12x - 4$$

$$= x^4 + 3x^3 - x^2 + 11x - 4$$

Note that the answer is written in descending order of the exponents, ending with the constant term.

EXAMPLE 10

Multiply: $(x + 2)(x^3 - 1)$.

Solution: $(x + 2)(x^3 - 1) = x^4 - x + 2x^3 - 2$

$$= x^4 + 2x^3 - x - 2$$

Unfortunately we do have situations where trinomials must be multiplied together. Method 3 can be extended to use in such a case, as Problem 7 illustrates.

Problem 7

Multiply: $(3x^2 + x + 6)(5x^3 - 3x + 1)$.

Solution: $3x^2 (5x^3 - 3x + 1) + x(5x^3 - 3x + 1) + 6(5x^3 - 3x + 1) =$

Answer: $15x^5 + 5x^4 + 21x^3 - 17x + 6$

Before completing the unit, let's look at several more examples that illustrate a "special product"—a binomial squared.

EXAMPLE 11

Multiply: $(x - 3)^2$.

Solution: $(x - 3)^2 = (x - 3)(x - 3)$

$$= x(x - 3) - 3(x - 3)$$
$$= x^2 - 3x - 3x + 9$$
$$= x^2 - 6x + 9$$

Note that the answer has the first and last terms squared but there is also a **middle term that is twice the product of the two terms** of the binomial.

$$(a + b)^2 = (a + b)(a + b)$$
$$= a(a + b) + b(a + b)$$
$$= a^2 + ab + ab + b^2$$
$$= a^2 + 2ab + b^2$$

This problem and Example 11 illustrate the following statement:

> When a binomial is squared, the result will be the first term squared plus twice the product of the two terms plus the last term squared.

$$(a + b)^2 = a^2 + 2ab + b^2$$

first term squared   twice the product of the terms   last term squared

---

EXAMPLE 12

Multiply: $(x + 7)^2$.

Solution: $(x + 7)^2 = x^2 + 2(7x) + 7^2$
$$= x^2 + 14x + 49$$

EXAMPLE 13

Multiply: $(x - 5)^2$.

Solution: $(x - 5)^2 = x^2 + 2(-5x) + 25$
$$= x^2 - 10x + 25$$

EXAMPLE 14

Multiply: $x(x + 4)^2$.

Solution: As always, remove parentheses first.
$$x(x + 4)^2 = x(x^2 + 2[4x] + 16)$$
$$= x(x^2 + 8x + 16)$$
$$= x^3 + 8x^2 + 16x$$

EXAMPLE 15

Multiply: $2x^3(x - 2)^2$.

Solution: $2x^3(x - 2)^2 = 2x^3(x^2 + 2[-2x] + 4)$
$$= 2x^3(x^2 - 4x + 4)$$
$$= 2x^5 - 8x^4 + 8x^3$$

You should now be able to multiply polynomials. Again I suggest using Method 3, which is similar to removing parentheses and will work for all cases.

Before beginning the next unit you should try to solve the following problems.

EXERCISES

In each case, perform the indicated multiplication. Try to do each problem in as few steps as possible.

1. $2cx^2(5c^2 - c - 3x)$
2. $(x + 4)(x + 5)$
3. $(x - 7)(x - 2)$
4. $(x - 1)(x - 5)$
5. $(x + 2)(x - 3)$
6. $(a + 5)^2$
7. $(x + 2)(x - 2)$
8. $(x - 1)^2$
9. $(2x + 3)(x - 1)$
10. $(5x - 2y)(3x + 7y)$
11. $(3x + 1)(2x - 5)$
12. $x(x - 4)^2$
13. $x(x - 5)^2$
14. $(x - 2)(x^3 - 4x^2 + 7x - 1)$
15. $(x^2 + 1)(x^2 - 3)$
16. $(x + 2y)(x - 3y)$
17. $(2a - 1)(3 - a - a^3)$
18. $(x^2 - 3x + 1)(x^3 - 2x)$
19. $(x^2 + 5)(x - 3)$
20. $2xy^2(x + y)(x - 3)$
21. $(5x - 2)(x + 2)$
22. $-3(x - 1)(x - 2)$
23. $(3a + b)(2a - 2b)$
24. $(5a - 3b)(-2a + 6b)$

If additional practice is needed:
  Leithold, page 65, problems 1–6, 9–20
  Lial and Miller, page 28, problems 27–40
  Peters, pages 35 and 36, problems 1–32, 49–72
  Rees, Sparks, and Rees, page 42, problems 9–40

UNIT 17

Division of Polynomials

In this unit we will examine the procedure for dividing polynomials. Although this kind of division does not occur very often, it is occasionally necessary to perform the operation in the course of solving a problem. The process outlined here will help on such occasions.

TWO IMPORTANT DEFINITIONS

Before studying the procedure for dividing polynomials, let's look at two definitions.

> Definition: A **polynomial** in x is said to be in **standard form** if:
>
> 1. All parentheses are removed.
> 2. Like terms are combined.
> 3. The terms are arranged in order of descending powers of x.

> Definition: The **degree of polynomial** in x is the highest power of x when the polynomial is written in standard form.

EXAMPLE 1

Write $x + 2x(x^2 - 5)$ in standard form and find its degree.

- -

Solution: $x + 2x(x^2 - 5) = x + 2x^3 - 10x$

$$= 2x^3 - 9x$$

Standard form: $2x^{\textcircled{3}} - 9x$

Degree: $3 \leftarrow$

- -

PROCEDURE FOR DIVIDING POLYNOMIALS

Now let's look at the procedure for dividing polynomials. Consider this problem:

Divide: $(5x^2 - 3x + 1)$ by $(x - 2)$.

The basic procedure is identical to the long-division process in arithmetic.

Step 1. Arrange both dividend and divisor in standard form and set up as a long-division problem, leaving spaces for any missing terms.

$$x - 2\overline{)\,5x^2 - 3x + 1}$$

Step 2. Divide the first term of the divisor into the first term of the dividend. (Recall that, since we are dividing, the exponents are subtracted.)

$$\begin{array}{r} 5x \\ x - 2\overline{)\,5x^2 - 3x + 1} \end{array}$$

Step 3. Multiply each term of the divisor by the first term of the quotient.

$$\begin{array}{r} 5x \\ x - 2\overline{)\,5x^2 - 3x + 1} \\ 5x^2 - 10x \end{array}$$

Step 4. Subtract like terms and bring down the next term from the dividend.

$$\begin{array}{r} 5x \\ x - 2\overline{)\,5x^2 - 3x + 1} \\ 5x^2 - 10x \downarrow \\ +7x + 1 \end{array}$$

Step 5. Repeat steps 2, 3, and 4, using the new remainder as the dividend.

$$\begin{array}{r} +7 \\ x - 2\overline{)} \\ 7x + 1 \\ 7x - 14 \\ \hline 15 \end{array}$$

Step 6. Continue repeating steps 2, 3, and 4 until the degree of the remainder is less than the degree of the divisor.

Therefore

$$(5x^2 - 3x + 1) \text{ divided by } (x - 2) \text{ equals } 5x + 7 + \frac{15}{x - 2}$$

Note that the remainder is written as a fraction.

Let me now try condensing some of the explanation. Keep in mind, though, that the procedure is identical to long division in arithmetic.

EXAMPLE 2

Divide: $(5x^2 + 7x - 1)$ by $(x + 3)$.

Solution: Step 2.   Divide first terms:

$$\frac{5x^2}{x} = 5x.$$

$$\begin{array}{r} 5x \\ x + 3 \overline{)\, 5x^2 + 7x - 1} \\ 5x^2 + 15x \downarrow \\ \hline -8x - 1 \end{array}$$

Step 3.   Multiply:

$$5x(x + 3) = 5x^2 + 15x.$$

Step 4.   Subtract; bring down -1.

Repeat steps 2, 3, and 4, using $-8x - 1$.

Step 2.   Divide first terms:

$$\frac{-8x}{x} = -8.$$

$$\begin{array}{r} 5x - 8 \\ x + 3 \overline{)\, 5x^2 + 7x - 1} \\ 5x^2 + 15x \\ \hline -8x - 1 \\ -8x - 24 \\ \hline 23 \end{array}$$

Step 3.   Multiply:

$$-8(x + 3) = -8x - 24.$$

Step 4.   Subtract.

The division process is finished because the degree of the remainder is less than the degree of the divisor.

Write the remainder as a fraction.

$$\begin{array}{r} 5x - 8 + \dfrac{23}{x + 3} \\ x + 3 \overline{)\, 5x^2 + 7x - 1} \\ 5x^2 + 15x \\ \hline -8x - 1 \\ -8x - 24 \\ \hline +23 \end{array}$$

Therefore:

$$(5x^2 + 7x - 1) \div (x + 3) = 5x - 8 + \frac{23}{x + 3}.$$

Here are two more examples. Example 3 is given in detail, with explanation.

EXAMPLE 3

Divide: $(2x^3 + 9x^2 + x + 5)$ by $(x^2 + 3x - 1)$.

Solution: Step 1. Set problem up.

Step 2. Divide first terms:

$$\frac{2x^3}{x^2} = 2x.$$

Step 3. Multiply:

$$2x(x^2 + 3x - 1).$$

Step 4. Subtract and bring down the next term of 5.

Repeat steps 2, 3, and 4, using $3x^2 + 3x + 5$.

$$
\require{enclose}
\begin{array}{r}
2x + 3 + \dfrac{-6x + 8}{x^2 + 3x - 1} \\[4pt]
x^2 + 3x - 1 \enclose{longdiv}{2x^3 + 9x^2 + x + 5} \\
\underline{2x^3 + 6x^2 - 2x} \downarrow \\
3x^2 + 3x + 5 \\
\underline{3x^2 + 9x - 3} \\
-6x + 8
\end{array}
$$

Step 2. Divide first terms:

$$\frac{3x^2}{x^2} = 3.$$

Step 3. Multiply:

$$3(x^2 + 3x - 1).$$

Step 4. Subtract.

The division process is finished because the degree of the remainder (1) is less than the degree of the divisor (2).

Write the remainder as a fraction.

$$\text{Answer:}\quad 2x + 3 + \frac{-6x + 8}{x^2 + 3x - 1}$$

EXAMPLE 4

Divide: $(x^4 + 3x^3 - x^2 + 11x - 4)$ by $(x + 4)$.

Solution:

$$
\require{enclose}
\begin{array}{r}
x^3 - x^2 + 3x - 1 \\
x + 4 \enclose{longdiv}{x^4 + 3x^3 - x^2 + 11x - 4} \\
\underline{x^4 + 4x^3} \\
-x^3 - x^2 \\
\underline{-x^3 - 4x^2} \\
3x^2 + 11x \\
\underline{3x^2 + 12x} \\
-x - 4 \\
\underline{-x - 4} \\
0
\end{array}
$$

$$\text{Answer:}\quad x^3 - x^2 + 3x - 1$$

You try this one.

Problem 1

Divide: $(x^3 - 4x^2 + 7x - 1)$ by $(x - 2)$.

Solution:

Answer: $x^2 - 2x + 3 + \dfrac{5}{x - 2}$

Look carefully at Examples 5–7.

EXAMPLE 5

Divide: $(x^3 + 2x + 5)$ by $(1 + x)$.

Solution:

$$
\begin{array}{r}
x^2 - x + 3 \\
x + 1 \overline{\smash{\big)}\ x^3 + 0\ + 2x + 5} \\
\underline{x^3 + x^2} \\
-x^2 + 2x \\
\underline{-x^2 - x} \\
3x + 5 \\
\underline{3x + 3} \\
2
\end{array}
$$

Answer: $x^2 - x + 3 + \dfrac{2}{x + 1}$

Note: The divisor had to be put in standard form, and a space had to be left for the missing x^2 term in the dividend. I inserted a zero in the space to avoid confusion when I subtracted.

EXAMPLE 6

Divide: $(7 - x + 12x^4)$ by $(2x^2 + 1)$.

Solution:

$$
\begin{array}{r}
6x^2 - 3 \\
2x^2 + 1 \overline{)\, 12x^4 + 0 \quad - x + 7} \\
\underline{12x^4 + 6x^2} \\
- 6x^2 - x + 7 \\
\underline{- 6x^2 - 3} \\
- x + 10
\end{array}
$$

Answer: $6x^2 - 3 + \dfrac{-x + 10}{2x^2 + 1}$

EXAMPLE 7

Divide: $(x^4 + 3)$ by $(x - 1)$.

Solution:

$$
\begin{array}{r}
x^3 + x^2 + x \ + \ 1 \\
x - 1 \overline{)\, x^4 + 0 + 0 + 0 + 3} \\
\underline{x^4 - x^3} \\
x^3 + 0 \\
\underline{x^3 - x^2} \\
x^2 + 0 \\
\underline{x^2 - x} \\
x + 3 \\
\underline{x - 1} \\
4
\end{array}
$$

Answer: $x^3 + x^2 + x + 1 + \dfrac{4}{x - 1}$

You should now be able to use the procedure for dividing polynomials. Remember that the procedure is identical to the long-division process in arithmetic.

Recall that, in short, the division procedure is:

1. **Set up** as long division.
2. **Divide** first terms.
3. **Multiply** the quotient times the divisor.
4. **Subtract** like terms; **bring down** the next term.
5. **Repeat,** using the new remainder as the dividend.
6. **Continue** until the degree of the remainder is less than the degree of the divisor.

Now try the following problems. Full solutions are given for all at the end of the book.

EXERCISES

Divide:

1. $(x^2 + 8x + 16) \div (x + 5)$
2. $(10x^2 - 13x + 1) \div (5x + 1)$
3. $(3x^4 + 2x^3 + 7x - 10) \div (x - 1)$
4. $(2 - 11x + 9x^2 + 6x^3) \div (2x^2 + 3x - 1)$
5. $(2x^3 + 3x^2 + x - 2) \div (2x - 3)$
6. $(x^3 + 2) \div (x + 2)$
7. $(3x^3 + x + 5 + 2x^2) \div (x^2 + x + 1)$

If additional practice is needed:

Leithold, pages 65 and 66, problems 39–53
Lial and Miller, pages 31 and 32, problems 1–16
Peters, pages 27 and 28, problems 1–8
Rich, page 70, problems 38–41

UNIT 18

Factoring Polynomials

In this unit we will discuss **factoring** of polynomials. Simply stated, factoring involves finding factors whose products are equal to the original polynomial.

We will look first at the simplest case, that of finding a common monomial factor. Then we will consider the factoring of trinomials.

FACTORING OUT A COMMON MONOMIAL FACTOR

First, recall the distributive law: $a(b + c) = ab + ac$.

multiplication $\longrightarrow$ $\longleftarrow$ factoring

Factoring out a common monomial factor is the application of the distributive law read from right to left.

A common monomial factor is a single expression that is a factor of each term of the polynomial.

Factoring out the common monomial factor is, in effect, taking out in front of the parentheses, *as much as possible*, what is common to all terms of the polynomial. It is just the reverse of removing parentheses.

Comment: The test of factoring is always whether you can remultiply and obtain the original polynomial.

EXAMPLE 1 $\qquad 6x^2y + 5x^2z = x^2(6y + 5z)$

EXAMPLE 2 $\qquad 2ay + 2 = 2(ay + 1)$

EXAMPLE 3 $\qquad 2xy - xy^2 + 3x^2y = xy(2 - y + 3x)$

EXAMPLE 4 $\qquad 8x^2 - 12x = 4x(2x - 3)$

EXAMPLE 5 $\qquad 12a^3 - 18a^2 + 3a = 3a(4a^2 - 6a + 1)$

Whenever you are factoring an expression, always first look to see whether there is a common monomial factor! Often the common monomial factor is overlooked, and the resulting problem is more difficult to handle.

FACTORING TRINOMIALS

Let us first consider factoring trinomials that are polynomials of degree 2 with the coefficient of x^2 being 1. In this unit such trinomials are factored experimentally—by trial and error—if they are factorable at all. However, the number of trials can often be reduced if certain observations are made.

We will classify trinomials into three categories, Cases I, II, and III, as follows.

Let D, E, a, and b be natural numbers.

Case I. $x^2 + Dx + E = (x + a)(x + b)$ where $ab = E$
$$a + b = D$$
all positive signs

Case I is the trinomial with all plus signs. This trinomial will factor into two binomials of the form $(x + a)(x + b)$ where a and b are numbers such that their product (ab) equals the constant of the trinomial and their sum $(a + b)$ equals the coefficient of the x term of the trinomial. The approach is based on applying the FOIL method in reverse.

EXAMPLE 6

Factor: $x^2 + 8x + 15$.

Solution: $x^2 + 8x + 15 = (x + ?)(x + ?)$

$x^2 + Dx + E = (x + a)(x + b)$ where $ab = E = 15$
$$a + b = D = 8$$

So we are looking for two numbers whose product is 15 and whose sum is 8. The numbers are 3 and 5.

Therefore:

$$x^2 + 8x + 15 = (x + 3)(x + 5)$$

EXAMPLE 7

Factor: $x^2 + 13x + 12$.

Solution: $x^2 + 13x + 12 = (x + ?)(x + ?)$

$x^2 + Dx + E = (x + a)(x + b)$ where $ab = E = 12$

$a + b = D = 13$

We need two numbers whose product is 12 and whose sum is 13. The numbers are 12 and 1.

Therefore:

$$x^2 + 13x + 12 = (x + 12)(x + 1)$$

EXAMPLE 8

Factor: $x^2 + 7x + 12$.

Solution: We need two numbers whose product is 12 and whose sum is 7. The numbers are 3 and 4.

Therefore:

$$x^2 + 7x + 12 = (x + 3)(x + 4)$$

The following are more examples of Case I. You might cover the answers and try factoring them yourself.

EXAMPLE 9 $x^2 + 10x + 9 = (x + 9)(x + 1)$

EXAMPLE 10 $x^2 + 5x + 6 = (x + 2)(x + 3)$

EXAMPLE 11 $x^2 + 13x + 22 = (x + 2)(x + 11)$

EXAMPLE 12 $x^2 + 8x + 12 = (x + 2)(x + 6)$

EXAMPLE 13 $x^2 + 12x + 35 = (x + 5)(x + 7)$

EXAMPLE 14 $x^2 + 7x + 1$ Prime—not factorable

Case II. $x^2 - Dx + E = (x - a)(x - b)$ where $ab = E$
 middle term only negative $a + b = D$

Case II is the trinomial in which the middle term *only* is negative. This trinomial will factor into two binomials of the form $(x - a)(x - b)$, where, as in Case I, a and b are numbers such that their product equals E and their sum equals D.

EXAMPLE 15

Factor: $x^2 - 9x + 14$.

Solution: $x^2 - 9x + 14 = (x - ?)(x - ?)$

$x^2 - Dx + E = (x - a)(x - b)$ where $ab = E = 14$
$a + b = D = 9$

We are looking for two numbers whose product is 14 and whose sum is 9 . The numbers are 7 and 2.

Therefore:

$$x^2 - 9x + 14 = (x - 2)(x - 7)$$

EXAMPLE 16

Factor: $x^2 - 7x + 12$.

Solution: $x^2 - 7x + 12 = (x - ?)(x - ?)$

$x^2 - Dx + E = (x - a)(x - b)$ where $ab = E = 12$
$a + b = D = 7$

The numbers are 3 and 4.

Therefore:

$$x^2 - 7x + 12 = (x - 3)(x - 4)$$

Case III. $x^2 \pm Dx - E = (x + a)(x - b)$ where $ab = E$

the difference of a and $b = D$

the larger of a and b has the sign of the middle term

last term negative

Case III is the trinomial with the *last* term negative. This will factor into two binomials of the form $(x + a)(x - b)$, where a and b are numbers such that, as before, their product equals E. But in this case the difference of a and b must equal D, and the larger of the two numbers has the same sign as the middle term of the trinomial.

I realize all that is a bit wordy. I hope Examples 17 and 18 will clarify any questions you may have.

EXAMPLE 17

Factor: $x^2 - 3x - 10$.

Solution: $x^2 - 3x - 10 = (x + ?)(x - ?)$

$x^2 - Dx - E = (x + a)(x - b)$ where $ab = E = 10$

the difference of a and b

$= D = 3$

We are looking for two numbers whose product is 10 and whose difference is 3. Obviously, the numbers are 2 and 5.

Since the middle term is negative, the 5 goes with the negative sign. Therefore:

$$x^2 - 3x - 10 = (x + 2)(x - 5)$$

Remember: the test of factoring is always whether you can multiply and obtain the original polynomial.

$$x^2 \qquad -10$$

Using the FOIL method: $(x + 2)(x - 5) = x^2 - 3x - 10$

and our factors
were correct.

$2x$
$-5x$

$-3x$

EXAMPLE 18

Factor: $x^2 + x - 20$.

Solution: $x^2 + x - 20 = (x + ?)(x - ?)$

$$x^2 + Dx - E = (x + a)(x - b) \qquad \text{where } ab = 20$$
$$\text{the difference of } a \text{ and } b$$
$$= D = 1$$

The numbers are 4 and 5. Since the middle term is positive, the 5 goes with the plus. Therefore:

$$x^2 + x - 20 = (x + 5)(x - 4)$$

The following seven problems are more examples of Case III for practice. Try factoring the trinomials yourself.

Problem 1 $x^2 - 4x - 21$

Problem 2 $x^2 + 5x - 6$

Problem 3 $x^2 + 6x - 16$

Problem 4 $x^2 - x - 56$

Problem 5 $x^2 + 2x - 15$

Problem 6 $x^2 + 8x - 9$

Problem 7 $x^2 - x - 7$

Answers: 1. $(x + 3)(x - 7)$ 2. $(x + 6)(x - 1)$ 3. $(x + 8)(x - 2)$
4. $(x + 7)(x - 8)$ 5. $(x + 5)(x - 3)$ 6. $(x + 9)(x - 1)$ 7. Prime

In reality, the three cases can be combined into one generalized case.

Generalized Case. $x^2 + Dx + E = (x + a)(x + b)$ where E is the product of a and b
D is the **algebraic** sum of a and b with D, E, a, and b integers

If we let D, E, a, and b be integers, the generalized case represents all three of the trinomials illustrated in Cases I, II, and III. This trinomial will factor into two binomials of the form $(x + a)(x + b)$, where a and b are integers such that their product equals E and their **algebraic** sum equals D.

If E is negative, a and b must have opposite signs because only the product of unlike signs yields a negative number.

If E is positive, a and b must have the same sign, either both positive or both negative.

Since D is the algebraic sum, its sign determines the sign of the numerically larger of a and b (whether their signs are the same or different).

Before beginning the next unit, try factoring the following polynomials.

EXERCISES

Factor

1. $x^2 + 2x - 3$
2. $x^2 - 15x + 56$
3. $x^2 + x$
4. $3x^2y - 12xy^2$
5. $3bx^2 + 27b^2$
6. $x^2 - x - 6$
7. $x^2 + 5x + 6$
8. $x^2 - 7x + 12$
9. $x^2 - x + 5$
10. $x^2 - 2x - 8$
11. $x^2 + 5x + 3$
12. $2x^3 + 2x^2 + 22x$
13. $5x^2 - 5x - 5$
14. $x^2 + 3x - 10$
15. $x^2 + 7x - 30$
16. $x^2 + 7x + 6$
17. $x^2 + 2x + 1$
18. $x^2 - 6x + 9$
19. $x^2 - x - 56$
20. $x^2 + 4x - 45$
21. $x^2 + 16x + 64$
22. $x^2 - 13x + 40$
23. $x^2 + 7x - 18$
24. $x^3 + x^2 + 5x$
25. $x^2 - 10x + 21$
26. $x^2 - 7x - 18$

If additional practice is needed:
   Drooyan and Wooton, Unit 4, page 60, exercise 2
   Peters, pages 40 and 41, problems 1–7, 41–51
   Rich, pages 179 and 181, problems 4, 19

UNIT 19

Factoring Special Binomials

In this unit we introduce some frequently occurring formulas that will aid you in factoring. Also, we consider a suggested procedure to follow when factoring any polynomial.

THREE IMPORTANT FORMULAS

There are certain formulas that occur frequently. It is helpful to memorize these formulas to facilitate recognition of them when factoring. These are formulas for the difference of two squares and for the sum or difference of two cubes.

$$\text{Difference of two squares:} \quad x^2 - y^2 = (x + y)(x - y)$$

The difference of two squares will factor into two binomials of the form $(x + y)(x - y)$, where x is the square root of the first term and y is the square root of the second term.

EXAMPLE 1

Factor: $a^2 - 25$.

Solution: The square root of a^2 is a.

The square root of 25 is 5.

Therefore:

$$a^2 - 25 = (a + 5)(a - 5)$$

The difference of two squares factors to the sum of their square roots times the difference of their square roots.

EXAMPLE 2 $\qquad x^2 - 1 = (x + 1)(x - 1)$

EXAMPLE 3 $\qquad 9x^2 - 4y^2 = (3x + 2y)(3x - 2y)$

EXAMPLE 4 $\qquad 16a^2 - 49 = (4a + 7)(4a - 7)$

EXAMPLE 5 $\qquad x^2y^2 - 100 = (xy + 10)(xy - 10)$

EXAMPLE 6 $\qquad y^2 + 25 \qquad$ Prime—not factorable. This is the sum of two squares.

EXAMPLE 7 $\qquad x^3 - 4 \qquad$ Prime—x^3 is not a square.

> Sum of two cubes: $\qquad x^3 + y^3 = (x + y)(x^2 - xy + y^2)$
>
> Difference of two cubes: $\quad x^3 - y^3 = (x - y)(x^2 + xy + y^2)$

Note the relationships in the two formulas above:

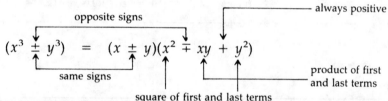

The sum of two cubes will factor into two polynomials of the form $(x + y)$ $(x^2 - xy + y^2)$, where x is the cube root of the first term and y is the cube root of the second term. A similar statement applies for the difference of two cubes.

EXAMPLE 8

Factor: $x^3 + 8$.

Solution: The cube root of x^3 is x.

The cube root of 8 is 2.

Therefore:

$$x^3 + y^3 = (x + y)(x^2 - xy + y^2)$$

$$x^3 + 8 = (x + 2)(x^2 - 2x + 4)$$

EXAMPLE 9

Factor: $27a^3 - 1$.

Solution: The cube root of $27a^3$ is $3a$.

The cube root of 1 is 1.

Therefore:

$$27a^3 - 1 = (3a - 1)(9a^2 + 3a + 1)$$

EXAMPLE 10

Factor: $64 - 27x^6$.

Solution: The cube root of 64 is 4.

The cube root of $27x^6$ is $3x^2$ because $(x^2)^3 = x^6$.

Therefore:

$$64 - 27x^6 = (4 - 3x^2)(16 + 12x^2 + 9x^4)$$

EXAMPLE 11

Factor: $125 + x^5$.

Solution: It is prime since x^5 is not a cube.

More often than not, to factor a polynomial completely requires more than one step. The following is a suggested procedure for factoring.

SUGGESTED PROCEDURE FOR FACTORING

1. Remove any common monomial factor.
2. If there are two terms, check for the difference of squares, difference of cubes, or sum of cubes, and factor accordingly.
3. If there are three terms, decide whether the trinomial is Case I, II, or III and try factoring the trinomial into two binomials.

EXAMPLE 12

Factor completely: $3x^4 - 27x^2$.

Solution:

common monomial factor
difference of two squares;
factor again

$$3x^4 - 27x^2 = 3x^2(x^2 - 9)$$
$$= 3x^2(x + 3)(x - 3)$$

EXAMPLE 13

Factor: $3x^4y - 6x^3y + 3x^2y$.

Solution:

common monomial factor
three terms—Case II;
factor again

$$3x^4y - 6x^3y + 3x^2y = 3x^2y(x^2 - 2x + 1)$$
$$= 3x^2y(x - 1)(x - 1)$$

EXAMPLE 14

Factor: $x^4 - 16$.

Solution:

difference of two squares;
factor again

$$x^4 - 16 = (x^2 + 4)(x^2 - 4)$$
$$= (x^2 + 4)(x + 2)(x - 2)$$

EXAMPLE 15

Factor: $2x^3 + 8x^2 + 6x$.

Solution:

common monomial factor

three terms—Case I;
factor again

$$2x^3 + 8x^2 + 6x = 2x(x^2 + 4x + 3)$$
$$= 2x(x + 1)(x + 3)$$

EXAMPLE 16

Factor: $(x + 3)^2 - 16$.

Solution: This is the difference of two squares.

The square root of $(x + 3)^2$ is $x + 3$.

The square root of 16 is 4.

Therefore:

$$(x + 3)^2 - 16 = [(x + 3) + 4][(x + 3) - 4]$$
$$= (x + 7)(x - 1)$$

EXAMPLE 17

Factor: $8x^3y^3 + 8y^3$.

Solution:

common monomial factor

two terms—sum of cubes;
factor again

$$8x^3y^3 + 8y^3 = 8y^3(x^3 + 1)$$
$$= 8y^3(x + 1)(x^2 - x + 1)$$

You should now have a good understanding of factoring and be able to factor many expressions. Remember our suggested three-step procedure which, simply stated, is as follows:

1.   Remove any common monomial factor.

2.   If there are two terms, check for difference of squares or sum or difference of cubes, and factor accordingly.

3.   If there are three terms, factor into two binomials, using Case I, II or III.

Before beginning the next unit, try factoring the following polynomials.

EXERCISES

Factor completely:

1. $5c^2 - 5c$
2. $m^2 + 10m + 25$
3. $x^2 - 9x - 10$
4. $2 - 2x^2$
5. $3b^2 - 75$
6. $9 - 6x + x^2$
7. $3abc^2 - 3abd^2$
8. $8x^3 + 27y^3$
9. $x^2 + 2x - 8$
10. $x^2 - x + 7$
11. $x^3 - 36x$
12. $x^7 - 8x^4y^3$
13. $x^2 + 13x + 30$
14. $3r^3 - 6r^2 - 45r$
15. $x^2 + 5x - 14$
16. $2a^2b^2c^2 - 4ab^2c^2 + 2b^2c^2$
17. $5x^2y - 15xy - 10y$
18. $5x^4 + 10x^3 - 15x^2$
19. $3x^2 - 12$
20. $a^2b^2 - a^2c^2$
21. $2xy^2 - 54xy + 100x$
22. $10ab^2 - 140ab + 330a$
23. $w^2x^2y^2 + 7w^2x^2y - 18w^2x^2$
24. $2ax^2 - 2ax - 40a$
25. $4a^2 - 9b^2$
26. $3a^2b - 3a^2b^5$
27. $4r^3s^2 - 48r^2s^2 + 108rs^2$
28. $2y^2z + 38yz + 96z$
29. $a^4 - b^4$
30. $a^2x^4 - 81a^2$

If additional practice is needed:
Drooyan and Wooton, Unit 4, page 62, exercise 4
Peters, pages 40 and 41, problems 8–28, 40–55
Rich, pages 181–183, problems 15, 27, 28

UNIT 20

Factoring (Continued)

This unit will continue our discussion of factoring. Specifically, we will learn a technique called **factoring by grouping**. Additionally, we will learn to factor trinomials in which the coefficient of the first (x^2) term is not 1.

FACTORING BY GROUPING

Recall from Unit 18 that we suggested this procedure for factoring:

1. Remove any common monomial factor.

2. If there are two terms, check for the difference of squares, difference of cubes, or sum of cubes, and factor accordingly.

3. If there are three terms, decide whether the trinomial is Case I, II, or III and try factoring the trinomial into two binomials.

   We will now add a fourth step:

4. If there are four terms, try grouping them into pairs that have some common variable.

EXAMPLE 1
Factor: $ab + ac - bd - cd$.

- -

Solution: Terms one and two have a in common.

Terms three and four have d in common and are both negative.
Therefore:
$$ab + ac - bd - cd = a(b + c) - d(b + c)$$

Since we can treat $(b + c)$ as a single quantity and $(b + c)$ is common to both terms, factor it out as a common factor:
$$= (b + c)(a - d)$$

Neat technique, isn't it?

- -

This technique is called **factoring by grouping**.

EXAMPLE 2

Factor: $2c - 6 - cy + 3y$.

Solution: $2c - 6 - cy + 3y = 2(c - 3) - y(c - 3)$

Since $(c - 3)$ is common to both terms, factor it out in front.

$= (c - 3)(2 - y)$

EXAMPLE 3

Factor: $2ax - 4bx + ay - 2by$.

Solution: $2ax - 4bx + ay - 2by = 2x(a - 2b) + y(a - 2b)$

$(a - 2b)$ is common; factor it out in front.

$= (a - 2b)(2x + y)$

EXAMPLE 4

Factor: $x^4 - x^3 + x - 1$.

Solution: $x^4 - x^3 + x - 1 = x^3(x - 1) + (x - 1)$

$(x - 1)$ is common; factor it out in front.

$= (x - 1)(x^3 + 1)$

Sum of cubes; factor again.

$= (x - 1)(x + 1)(x^2 - x + 1)$

It's your turn to try a problem.

Problem 1

Factor: $2axy + 4ay - 3x - 6$.

Solution:

Answer: $(x + 2)(2ay - 3)$

FACTORING TRINOMIALS IN WHICH THE COEFFICIENT OF x^2 IS NOT 1

When the coefficient of x^2 is not 1, the problem of factoring trinomials becomes far more complicated; but the procedure is similar to that of Unit 18.

EXAMPLE 5

Factor: $3x^2 + 4x + 1$.

Solution: Since there are only plus signs, the trinomial is similar to Case I and the two binomials will both have plus signs.

$$3x^2 + 4x + 1 = (\quad + \quad)(\quad + \quad)$$

Recall from the FOIL method diagram that:
the product of the first terms must be $3x^2$,
the product of the last terms must be 1, and
the sum of inner product and outer product must be $4x$.

To get $3x^2$ the factors are $3x$ and x.

To get 1 the factors are 1 and 1.

That leaves us with the possible binomial factors being $(3x + 1)(x + 1)$.

Checking to see whether the middle term is correct:

$$\begin{array}{r} x \\ +3x \\ \hline 4x \end{array}$$

Hence:

$$3x^2 + 4x + 1 = (3x + 1)(x + 1)$$

Remember that we can always verify our answer by multiplying; we should obtain the original polynomial.

EXAMPLE 6

Factor: $2x^2 - 5x + 3$.

Solution: Since the middle term only is negative, the trinomial is similar to Case II and the binomial factors will both have negative signs.

$$2x^2 - 5x + 3 = (\quad - \quad)(\quad - \quad)$$

To get $2x^2$ the factors are $2x$ and x.

To get 3 the factors are 1 and 3.

That leaves us with several choices.

The possible binomial factors are either

$(2x - 1)(x - 3)$ or $(2x - 3)(x - 1)$

$-x$ $-3x$
$-6x$ $-2x$
$-7x$ $-5x$

We try each one to see which has the correct middle term. Therefore:

$$2x^2 - 5x + 3 = (2x - 3)(x - 1)$$

EXAMPLE 7

Factor: $6x^2 - x - 15$.

Solution: Since the last term is negative, the trinomial is similar to Case III and the binomial factors will have different signs.

$$6x^2 - x - 15 = (\quad + \quad)(\quad - \quad)$$

To get $6x^2$ the possible factors are:

$(6x \quad)(x \quad)$ or $(2x \quad)(3x \quad)$

Possible factors for 15 are $15 \cdot 1$ and $5 \cdot 3$. So the possibilities are:

$(6x + 1)(x - 15)$	$(6x - 1)(x + 15)$	$(3x + 3)(2x - 5)$
$(6x + 15)(x - 1)$	$(6x - 15)(x + 1)$	$(3x - 3)(2x + 5)$
$(6x + 3)(x - 5)$	$(6x - 3)(x + 5)$	$(3x + 5)(2x - 3)$
$(6x + 5)(x - 3)$	$(6x - 5)(x + 3)$	$(3x - 5)(2x + 3)$
$(3x + 1)(2x - 15)$	$(3x - 1)(2x + 15)$	
$(3x + 15)(2x - 1)$	$(3x - 15)(2x + 1)$	

Fortunately, eight of the possibilities can be eliminated simply by observing that, since there is no common factor in the original trinomial, there can be none in any of its factored forms. For example, $(6x + 15)$ is impossible as a factor since $(6x + 15) = 3(2x + 5)$, but the original trinomial does not have a common factor of 3.

With this observation the list of binomial factors to be considered can be shortened to:

$$(6x + 1)(x - 15) \qquad (6x - 1)(x + 15) \qquad (3x + 5)(2x - 3)$$
$$(6x + 5)(x - 3) \qquad (6x - 5)(x + 3) \qquad (3x - 5)(2x + 3)$$
$$(3x + 1)(2x - 15) \qquad (3x - 1)(2x + 15)$$

Upon inspection we find that the binomial factors are:

$$(3x - 5)(2x + 3)$$
$$-10x$$
$$9x$$
$$-x$$

Therefore:

$$6x^2 - x - 15 = (3x - 5)(2x + 3)$$

___

This required quite a bit of writing, and any shortcut you may devise helps! The following example demonstrates a technique that represents one of many possible shortcuts. I will use the same trinomial as in Example 7 to illustrate the simplicity of the new technique.

EXAMPLE 8

Factor: $6x^2 - x - 15$.

$$6x^2 - x - 15$$

Solution: We look for two **integers** whose sum is -1 and whose product is $6(-15) = -90$. The numbers are 9 and -10.

Using the two numbers just found, we rewrite the original trinomial and then use **factoring by grouping**.

the two integers

$$6x^2 - x - 15 = 6x^2 + 9x - 10x - 15$$
$$= 3x(2x + 3) - 5(2x + 3)$$
$$= (2x + 3)(3x - 5)$$

___

EXAMPLE 9

Factor: $8x^2 + 26x + 15$.

$$8x^2 + 26x + 15$$

Solution: We look for two **integers** whose sum is 26 and whose product is $8(15) = 120$. These numbers are 20 and 6.
Rewrite the trinomial, using 20 and 6 as coefficients of x.
We will now use factoring by grouping to do the problem.

$$8x^2 + 26x + 15 = 8x^2 + 20x + 6x + 15$$

$$= 4x(2x + 5) + 3(2x + 5)$$

$$= (2x + 5)(4x + 3)$$

EXAMPLE 10

Factor: $12x^2 + 28x - 5$.

$$12x^2 + 28x - 5$$

Solution: We look for two **integers** whose sum is 28 and whose product is $12(-5) = -60$. The numbers are $+30$ and -2. Therefore:

$$12x^2 + 28x - 5 = 12x^2 + 30x - 2x - 5$$

$$= 6x(2x + 5) - (2x + 5)$$

$$= (2x + 5)(6x - 1)$$

We can generalize the situation, give $ax^2 + bx + c$ to factor, where a, b, and c are real numbers, and a is not equal to 1, as follows:

1. Look for two integers whose sum is b and whose product is ac.

2. Rewrite $ax^2 + bc + c$, using the integers for the coefficients of x, and complete the problem by factoring by grouping.

Do Problems 2 and 3, applying what you have learned in this unit.

Problem 2

Factor: $3x^2 + 13x + 4$.

Solution: We are looking for two integers whose sum is _____ and whose product is (   )(   ) = _____ .

The numbers are _____ and _____ . Therefore:

$$3x^2 + 13x + 4 = 3x^2 \underline{\hspace{6cm}} + 4$$

$$=$$

$$=$$

Answer: $(3x + 1)(x + 4)$

Problem 3

Factor: $2x^2 + 7x - 15$.

Solution:

Answer: $(2x - 3)(x + 5)$

Before going on to the next unit, try factoring the following expressions.

EXERCISES

Factor completely:

1. $7x^2 + 10x + 3$

2. $2y^2 + 5y - 3$

3. $7x^2 - 10x + 3$

4. $3x^3 - 5x^2 - 9x + 15$

5. $4x^2 + 28xy + 49y^2$

6. $6x^2 + 13x + 6$

7. $24 - 2a - 2a^2$

8. $4x^3 - 10x^2 - 6x + 9$

9. $5x^2 - 4x - 1$

10. $4x^2 + 8x + 4$

11. $7x^2 + 13x - 2$

12. $2x^2 - 7x + 6$

13. $2y^2 - 17y + 35$

14. $7x^2 + 32x - 15$

15. $6xw^2 + 16xw - 6x$

16. $6z^2 + 2z - 4$

17. $x^2y^2z - 16xyz + 64z$

18. $15x^2 - 2xy - 24y^2$

19. $6x^2 - 5x - 6$

20. $2x^2 + 5x - 2$

21. $8x^2 + 30x - 27$

22. $xy^3 + 2y^2 - xy - 2$

23. $12x^2 - 4x - 5$

24. $x^4 - y^4$

25. $1 - a^4$

If additional practice is needed:
  Drooyan and Wooton, Unit 4, page 62, exercise 5
  Peters, pages 41, problems 46–90
  Rees, Sparks, and Rees, pages 58 and 59, problems 1–68
  Rich, pages 182, problems 20, 21

UNIT 21

Solving Second-Degree Equations—Factoring

> Definition: $ax^2 + bx + c = 0$, with a, b, and c being real numbers, $a \neq 0$, is called a **second-degree equation** or **quadratic equation**.

In other words, a second-degree equation must contain a squared term, x^2, and no higher powered term.

Examples of second-degree equations are:

$$5x^2 + 1 = 0$$
$$x^2 = 25$$
$$x^2 - 5x + 1 = 7$$
$$2x + x^2 = 15x$$

whereas

$$x^3 + 2x = 5 \text{ is third degree}$$
$$x^4 = 0 \text{ is fourth degree}$$
$$x^5 + 3x^2 - 5x + 27 = 7x^3 \text{ is fifth degree}$$

In this unit we will deal only with second-degree equations, saving the others for Unit 23.

To solve a second-degree or quadratic equation is to find the values of x that satisfy the equation. With a second-degree equation there will be **at most two real solutions**, "at most" meaning there could be two unequal real solutions, one solution from two equal real solutions resulting in one value, or no real solution.

There are basically three ways to solve quadratic equations:

1. Factoring—is the fastest method, but not always possible.

2. Completing the square—is cumbersome and lengthy and so will not be developed in this book.

3. Quadratic formula—will always work, but is long!

Definition: $ax^2 + bx + c = 0$, is called the **standard form** of the quadratic equation $(a \neq 0)$.

Note: **All** terms are on the left side of the equal sign, with **only** 0 on the right.

The technique I will use involves four steps:

1. Write the equation in standard form. In other words, put **all** terms on the left side, with **only** 0 remaining on the right side of the equal sign.

2. Factor the left side of the equation.

3. Set **each** factor equal to 0.

4. Solve the new equations from step 3.

EXAMPLE 1

Solve: $x^2 - 3x = -2$.

Solution: $$x^2 - 3x = -2$$

Step 1. $x^2 - 3x + 2 = 0$

Step 2. $(x - 1)(x - 2) = 0$

Step 3. $x - 1 = 0$ or $x - 2 = 0$

$$x = 1 \qquad\qquad x = 2$$

Comment: Notice that we have **two** solutions. Both must satisfy the original equation. To verify, check by substitution.

$$x^2 - 3x = -2$$

If $x = 1$: $(1)^2 - 3(1) \overset{?}{=} -2$

$1 - 3 \overset{?}{=} -2$

$-2 = -2$

If $x = 2$: $(2)^2 - 3(2) \overset{?}{=} -2$

$4 - 6 \overset{?}{=} -2$

$-2 = -2$

EXAMPLE 2

Solve: $x^2 = -6 - 5x$.

Solution: $x^2 = -6 - 5x$

        Step 1. $x^2 + 5x + 6 = 0$

        Step 2. $(x + 2)(x + 3) = 0$

        Step 3. $x + 2 = 0$ or $x + 3 = 0$

                     $x = -2$ $x = -3$

Here are two problems for you.

Problem 1

Solve: $x^2 - 2x - 48 = 0$.

Solution:

Answer: $x = 8$ or $x = -6$

Problem 2

Solve: $x^2 = 5x - 4$.

Solution:

Answer: $x = 1$ or $x = 4$

SOLVING SECOND-DEGREE EQUATIONS—FACTORING **141**

Now let's try several more examples, using the same technique with slight variations.

EXAMPLE 3

Solve: $-x^2 + 2x + 3 = 0$.

Solution: $-x^2 + 2x + 3 = 0$

$x^2 - 2x - 3 = 0$ Note: Multiplying the entire equation by -1 makes the factoring easier.

Finish the problem:

Answer: $x = 3$ or $x = -1$

EXAMPLE 4

Solve: $x^2 - 7x + 6 = -4$.

Solution: $x^2 - 7x + 6 = -4$

$x^2 - 7x + 10 = 0$

$(x - 2)(x - 5) = 0$

$x - 2 = 0$ or $x - 5 = 0$

$x = 2$ $x = 5$

EXAMPLE 5

Solve: $x^2 - 4 = 6x - 13$.

Solution: $x^2 - 4 = 6x - 13$

Step 1. $x^2 - 6x + 9 = 0$

Step 2. $(x - 3)(x - 3) = 0$

Step 3. $x - 3 = 0$ or $x - 3 = 0$

Step 4. $x = 3$ $x = 3$

There is only one solution since the two factors in step 2 are the same, resulting in two equal roots.

EXAMPLE 6

Solve: $x^2 - 25 = 0$.

Solution: $x^2 - 25 = 0$

$(x - 5)(x + 5) = 0$

$x - 5 = 0$ or $x + 5 = 0$

$x = 5$ $x = -5$

Alternative solution:

$x^2 - 25 = 0$

$x^2 = 25$

$x = \pm 5$

Since there is no x term, take the square root of both sides. Be sure to write both answers, plus and minus:

$x = 5$ or $x = -5$.

EXAMPLE 7

Solve: $x^2 + 4 = 0$.

Solution: $x^2 + 4 = 0$

$x^2 + 4$ is prime and cannot be factored; hence there is no solution.

Alternative solution: $x^2 + 4 = 0$

Since there is no x term: $x^2 = -4$

There is no real solution since there is no real number squared that equals a negative number.

For some reason most students have difficulty in factoring the next example, where there is no constant term. In reality it is the easiest to factor!

EXAMPLE 8

Solve: $x^2 - 3x = 0$.

Solution: $x^2 - 3x = 0$

$x(x - 3) = 0$

$x = 0$ or $x - 3 = 0$

$x = 0$ $x = 3$

So there are two solutions.

EXAMPLE 9

Solve: $2x^2 - x = 0$.

Solution:
$$2x^2 - x = 0$$
$$x(2x - 1) = 0$$
$$x = 0 \quad \text{or} \quad 2x - 1 = 0$$
$$2x = 1$$
$$x = \tfrac{1}{2}$$

The two solutions are $x = 0$ or $x = \tfrac{1}{2}$.

If you have difficulty factoring the next two examples, don't worry. In Unit 22 I will explain the quadratic formula that allows you to solve a second-degree equation without any factoring.

EXAMPLE 10

Solve: $3x^2 = x + 2$.

Solution:
$$3x^2 = x + 2$$
$$3x^2 - x - 2 = 0$$
$$(3x + 2)(x - 1) = 0$$
$$3x + 2 = 0 \quad \text{or} \quad x - 1 = 0$$
$$3x = -2 \qquad\qquad x = 1$$
$$x = \frac{-2}{3}$$

EXAMPLE 11

Solve: $5x - 3 = -2x^2$.

Solution:
$$5x - 3 = -2x^2$$
$$2x^2 + 5x - 3 = 0$$
$$(2x - 1)(x + 3) = 0$$
$$2x - 1 = 0 \quad \text{or} \quad x + 3 = 0$$
$$2x = 1 \qquad\qquad x = -3$$
$$x = \tfrac{1}{2}$$

To summarize, in Unit 21 we defined, and you should now be able to identify, a second-degree equation as of the type $ax^2 + bx + c = 0$ with $a \neq 0$. Also, you should expect at most two different real solutions. And to find the solutions you should:

1.   Write the equation in standard form.

2.   Factor the left side.

3.   Set each factor equal to 0.

4.   Solve the new equations from step 3.

Before beginning the next unit you should solve the following equations.

EXERCISES

Solve for x:

1. $x^2 + 5x - 14 = 0$

2. $x^2 + 13x + 30 = 0$

3. $x^2 - x + 7 = 0$

4. $4x^2 + 8x + 4 = 0$

5. $x^2 + 5x = 0$

6. $x^2 + 2x = 8$

7. $2t^2 + 20t + 50 = 0$

8. $5x^2 - 5x = 0$

9. $2x^2 + 6 = 7x$

10. $2x^2 + 8x + 6 = 0$

11. $z^2 + 4z - 21 = 0$

12. $10x - 10 = 19x - x^2$

13. $3x^2 + 2x = 0$

14. $2 - 2x^2 = 0$

15. $2w^2 + 7w - 4 = 0$

16. $2x^2 + 7x - 15 = 0$

17. $10a^2 - a - 3 = 0$

18. $2x^2 - 5x - 3 = 0$

19. $81x^2 - 144 = 0$

20. $12x^2 + 5x - 2 = 0$

If additional practice is needed:
   Drooyan and Wooton, Unit 8, page 81, exercise 1
   Leithold, page 161, problems 1–18
   Peters, page 254, problems 1–20
   Rees, Sparks, and Rees, pages 180, problems 1–40

UNIT 22

Solving Second-Degree Equations—Quadratic Formula

In this unit we will discuss the use of the quadratic formula for solving second-degree equations that are not readily factored.

Quadratic Formula

Given a quadratic equation $ax^2 + bx + c = 0$, with $a \neq 0$, the solutions are given by

$$x = \frac{-b \pm \sqrt{b^2 - 4ac}}{2a}$$

Comments: Consider the number under the radical, $b^2 - 4ac$:

1. If $b^2 - 4ac$ is negative, there is no real solution since we cannot take the square root of a negative number in the reals.

2. If $b^2 - 4ac = 0$, there are two real and equal solutions.

3. If $b^2 - 4ac$ is positive, there are two solutions. One is found by using the plus sign, and the other with the minus sign.

Now our technique for solving second-degree equations has been revised to:

1. Write the equation in standard form.

2. Factor the left-hand side. (If unable to factor, use the quadratic formula after determining the values for a, b, and c).

Let us assume for the remainder of the unit that you are terrible at factoring and must resort to the quadratic formula at all times!

EXAMPLE 1

Solve: $3x^2 = x + 2$.

Solution:

$$3x^2 = x + 2$$

Compare with:

$$3\,x^2 - x - 2 = 0$$
$$a\,x^2 + b\,x + c = 0$$

a is the coefficient of the squared term,

b is the coefficient of the first-degree term,

c is the constant,

when the **equation is in standard form**.

Thus $a = 3$, $b = -1$, $c = -2$.

Now, using the formula and substituting:

$$x = \frac{-b \pm \sqrt{b^2 - 4ac}}{2a}$$

$$= \frac{-(-1) \pm \sqrt{1 - 4(-6)}}{6} \quad \text{since } b^2 = (-1)^2 = 1,$$
$$ac = 3(-2) = -6$$

$$= \frac{1 \pm \sqrt{1 + 24}}{6}$$

$$= \frac{1 \pm \sqrt{25}}{6}$$

$$= \frac{1 \pm 5}{6} \quad \text{since } \sqrt{25} = 5$$

Now, to find the two solutions, use first the plus and then the minus:

$$x = \frac{1 + 5}{6} \quad \text{or} \quad x = \frac{1 - 5}{6}$$

$$= \frac{6}{6} \qquad\qquad = \frac{-4}{6}$$

$$= 1 \qquad\qquad\quad = \frac{-2}{3}$$

Note: Had we factored $3x^2 - x - 2 = (x - 1)(3x + 2)$, we would have obtained the same solution.

EXAMPLE 2

Solve: $-5x = -3x^2 + 12$.

Solution:

$$-5x = -3x^2 + 12$$

Compare with: $\underbrace{3}_{a}x^2 \underbrace{- 5}_{+ b}x \underbrace{- 12}_{+ c} = 0$

$$a x^2 + b x + c = 0$$

Thus $a = 3$, $b = -5$, $c = -12$.

Substitute into the quadratic formula:

$$x = \frac{-b \pm \sqrt{b^2 - 4ac}}{2a}$$

$$= \frac{-(-5) \pm \sqrt{25 - 4(-36)}}{2(3)}$$ since $b^2 = (-5)^2 = 25$,
$$ac = 3(-12) = 36$$

$$= \frac{5 \pm \sqrt{25 + 144}}{6}$$

$$= \frac{5 \pm \sqrt{169}}{6}$$

$$= \frac{5 \pm 13}{6}$$ since $\sqrt{169} = 13$

The two solutions are given by:

$$x = \frac{5 + 13}{6} \qquad \text{or} \qquad x = \frac{5 - 13}{6}$$

$$= \frac{18}{6} \qquad\qquad\qquad = \frac{-8}{6}$$

$$= 3 \qquad\qquad\qquad = \frac{-4}{3}$$

Note: Factoring:

$$3x^2 - 5x - 12 = 0$$

$$(x - 3)(3x + 4) = 0$$

would have led to the same result.

Are you ready to try one?

Problem 1

Solve: $2x^2 = 5x + 7$.

Answer: $x = -1$ or $x = \frac{7}{2}$

Now study Examples 3–6.

EXAMPLE 3

Solve: $x^2 - 2x + 4 = 0$.

Solution:

$$x^2 - 2x + 4 = 0$$

Compare: $ax^2 + bx + c = 0$

Thus $a = 1, b = -2, c = 4$.

Substitute into the quadratic formula:

$$x = \frac{-b \pm \sqrt{b^2 - 4ac}}{2a}$$

$$= \frac{-(-2) \pm \sqrt{4 - 4(4)}}{2} \qquad \text{since } b^2 = (-2)^2 = 4,$$
$$ac = 1(4) \quad = 4$$

$$= \frac{2 \pm \sqrt{4 - 16}}{2}$$

$$= \frac{2 \pm \sqrt{-12}}{2}$$

We can stop right here and write:

No real solution, since we cannot have a negative number under the square root symbol in the reals.

EXAMPLE 4

Solve: $4x^2 - 4x + 1 = 0$.

Solution:

$$\text{Compare:} \quad \begin{array}{ccc} \boxed{4}x^2 & \boxed{-4}x & \boxed{+1} = 0 \\ \boxed{a}x^2 & +\boxed{b}x & +\boxed{c} = 0 \end{array}$$

Thus $a = 4$, $b = -4$, $c = 1$.

Substitute into the quadratic formula:

$$x = \frac{-b \pm \sqrt{b^2 - 4ac}}{2a}$$

$$= \frac{-(-4) \pm \sqrt{16 - 4(4)}}{2(4)} \qquad \text{since } b^2 = (-4)^2 = 16,$$
$$ac = 4(1) \quad = 4$$

$$= \frac{4 \pm \sqrt{16 - 16}}{8}$$

$$= \frac{4 \pm \sqrt{0}}{8} \qquad \text{since the square root of 0 is 0}$$

$$= \frac{4}{8}$$

$$= \frac{1}{2}$$

There are two real and *equal* solutions.

EXAMPLE 5

Solve: $\frac{1}{3}x^2 - 3x - 2 = 0$.

Solution: Recall from **Unit 3** that, whenever we attempted to solve an equation containing fractions, the first step was to simplify. That means to multiply the entire equation by the common denominator of all fractions.

Following that advice with this example, we should first multiply the entire equation by 3 to clear of fractions.

$$3(\tfrac{1}{3}x^2 - 3x - 2 = 0)$$

$$x^2 - 9x - 6 = 0$$

Compare: $ax^2 + bx + c = 0$

Therefore $a = 1$, $b = -9$, $c = -6$.

Substitute into the quadratic equation:

$$x = \frac{-b \pm \sqrt{b^2 - 4ac}}{2a}$$

$$= \frac{-(-9) \pm \sqrt{81 - 4(-6)}}{2(1)} \qquad \text{since } b^2 = (-9)^2 = 81,$$
$$ac = 1(-6) = -6$$

$$= \frac{9 \pm \sqrt{81 + 24}}{2}$$

$$= \frac{9 \pm \sqrt{105}}{2}$$

Or, if you prefer,

$$x = \frac{9 + \sqrt{105}}{2} \qquad \text{or} \qquad x = \frac{9 - \sqrt{105}}{2}$$

Note: $(9 \pm \sqrt{105})/2$ are **exact** answers to the equation. Any attempt to obtain a decimal equivalent for $\sqrt{105}$ by use of a calculator or table will result only in the approximation because $\sqrt{105}$ is irrational.

The application to be made of the answer determines which form should be used.

EXAMPLE 6

Solve: $3x^2 + x = 6$.

Solution:

$$3x^2 + x - 6 = 0$$

Compare: $ax^2 + bx + c = 0$

Thus $a = 3$, $b = 1$, $c = -6$.

Substitute:

$$x = \frac{-b \pm \sqrt{b^2 - 4ac}}{2a}$$

$$= \frac{-1 \pm \sqrt{1 - 4(-18)}}{2(3)}$$

$$= \frac{-1 \pm \sqrt{1 + 72}}{6}$$

$$= \frac{-1 \pm \sqrt{73}}{6}$$

Or, if you prefer,

$$x = \frac{-1 + \sqrt{73}}{6} \quad \text{or} \quad x = \frac{-1 - \sqrt{73}}{6}$$

It's your turn again.

Problem 2

Solve: $3x^2 + x - 1 = 0$.

Answer: $x = \dfrac{-1 \pm \sqrt{13}}{6}$

Here are two more examples to end the unit.

EXAMPLE 7

Solve: $x^2 - 4x - 7 = 0$.

Solution: Skipping a few steps and going right to the quadratic formula with $a = 1$, $b = -4$, $c = -7$:

$$x = \frac{-b \pm \sqrt{b^2 - 4ac}}{2a}$$

$$= \frac{-(-4) \pm \sqrt{16 - 4(-7)}}{2(1)}$$

$$= \frac{4 \pm \sqrt{16 + 28}}{2}$$

$$= \frac{4 \pm \sqrt{44}}{2}$$

At this point you have three choices:

1.   Stop and leave the answer as above.

2.   Use tables or calculator to convert to decimals.

3.   Simplify the answer further as follows:

$$x = \frac{4 \pm \sqrt{4 \cdot 11}}{2}$$

$$= \frac{4 \pm 2\sqrt{11}}{2} \quad \text{since } \sqrt{4} = 2$$

$$= \frac{2(2 \pm \sqrt{11})}{2} \quad \text{by factoring out 2}$$

$$= \frac{\cancel{2}(2 \pm \sqrt{11})}{\cancel{2}} \quad \text{and cancelling}$$

$$= 2 \pm \sqrt{11} \quad \text{is the simplified answer.}$$

EXAMPLE 8

Solve: $2x^2 - 2x = 1$.

Solution:

$$2x^2 - 2x = 1$$

$$2x^2 - 2x - 1 = 0$$

Compare: $ax^2 + bx + c = 0$

Thus $a = 2$, $b = -2$, $c = -1$.

Substitute into the quadratic formula:

$$x = \frac{-b \pm \sqrt{b^2 - 4ac}}{2a}$$

$$= \frac{-(-2) \pm \sqrt{4 - 4(-2)}}{2(2)}$$

$$= \frac{2 \pm \sqrt{4 + 8}}{4}$$

$$= \frac{2 \pm \sqrt{12}}{4}$$

To further simplify the answer, if necessary, proceed as follows:

$$x = \frac{2 \pm \sqrt{4 \cdot 3}}{4} \quad \text{since } 12 = 4 \cdot 3$$

$$= \frac{2 \pm 2\sqrt{3}}{4} \quad \text{since } \sqrt{4} = 2$$

$$= \frac{2(1 \pm \sqrt{3})}{4} \quad \text{by factoring out 2}$$

$$= \frac{\cancel{2}(1 \pm \sqrt{3})}{\cancel{2} \cdot 2} \quad \text{and cancelling}$$

$$= \frac{1 \pm \sqrt{3}}{2}$$

You now should be able to solve any second-degree (quadratic) equation. The procedure is to write the equation to be solved in standard form. Then, if the left side cannot be readily factored, compare it with $ax^2 + bx + c = 0$ to determine the values of a, b, and c. These values are then substituted into the quadratic formula

$$x = \frac{-b \pm \sqrt{b^2 - 4ac}}{2a}$$

Now try to solve the following equations.

EXERCISES

Solve for x:

1. $x^2 + 3x - 1 = 0$

2. $2x^2 - 3x - 2 = 0$

3. $4x^2 + 4x - 4 = 0$

4. $2x^2 - 3x - 1 = 0$

5. $x^2 + x - 20 = 0$

6. $\frac{1}{5}x^2 - 5x + 1 = 0$

7. $\frac{4}{x} = x - 3, \quad x \neq 0$

8. $x^2 - 2x + 2 = 0$

9. $2x^2 - x = 0$

10. $4x^2 - 4x + 1 = 0$

11. $2x^2 + 7x + 9 = 0$

12. $x^2 - 2x - 10 = 0$

13. $3x^2 + x - 3 = 0$

14. $2x^2 + \sqrt{3}x - 4 = 0$

_ _

If additional practice is needed:
   Drooyan and Wooton, Unit 8, page 83, exercise 5
   Leithold, page 170, problems 1–26
   Peters, page 259, problems 1–10
   Rees, Sparks, and Rees, pages 188, problems 1–16
   Rich, page 233, problem 15

UNIT 23

Solving Third-Degree and Higher Equations

The objective of this unit is to illustrate the approach used to solve third-degree and higher equations. The technique is basically the same as that presented in Unit 21.

Step 1. Bring all the terms to the left-hand side, with only 0 remaining on the right of the equal sign.

Step 2. Factor the left-hand side.

Step 3. Set each factor equal to 0.

Step 4. Solve the new equations from step 3.

Unfortunately, if we are unable to factor the expression, there is *no such thing* as the quadratic formula to help us out.

Recall that with a second-degree equation there are at most two real and different solutions. Similarly, with a third-degree equation there will be at most three solutions, with a fourth-degree equation at most four solutions, and so on.

EXAMPLE 1

Solve: $x^3 - 7x^2 + 6x = 0$.

Solution: Step 1. $x^3 - 7x^2 + 6x = 0$

Factor out an x: Step 2. $x(x^2 - 7x + 6) = 0$

Factor again: Step 2. $x(x - 1)(x - 6) = 0$

Step 3. $x = 0$, $x - 1 = 0$, $x - 6 = 0$

Step 4. $x = 1$ $x = 6$

These are the three solutions:

$$x = 0, \qquad x = 1, \qquad x = 6$$

You can verify that each one of these is indeed a solution by substituting it back into the original equation.

EXAMPLE 2

Solve: $x^4 - 16 = 0$.

Solution: $x^4 - 16 = 0$

$$(x^2 - 4)(x^2 + 4) = 0$$

$$(x - 2)(x + 2)(x^2 + 4) = 0$$

$x - 2 = 0$, $x + 2 = 0$, $x^2 + 4 = 0$

$x = 2$ $x = -2$ $x^2 = -4$

no real solution
to this part

Thus there are only two solutions to this fourth-degree equation:

$$x = 2, \qquad x = -2.$$

EXAMPLE 3

Solve: $x^3 - x = 0$.

Solution: $x^3 - x = 0$

$$x(x^2 - 1) = 0$$

$$x(x - 1)(x + 1) = 0$$

$x = 0$, $x - 1 = 0$, $x + 1 = 0$

$x = 1$ $x = -1$

The three solutions are:

$$x = 0, \qquad x = 1, \qquad x = -1.$$

Now you try a problem.

Problem 1

Solve: $x^3 - 3x^2 + 2x = 0$.

Solution:

Answer: $x = 0, 1, 2$

I'll do two more examples.

EXAMPLE 4

Solve: $(x + 1)(x - 3)(x + 4) = 0$.

Solution: Most of the work has been done for us because the equation is already factored.

$$(x + 1)(x - 3)(x + 4) = 0$$

$$x + 1 = 0, \qquad x - 3 = 0, \qquad x + 4 = 0$$

$$x = -1 \qquad x = 3 \qquad x = -4$$

EXAMPLE 5

Solve: $2x(2x + 1)(x - 2)(3x - 9) = 0$.

Solution: Again, since the entire left side has been factored, we can move immediately to step 3:

$$2x(2x + 1)(x - 2)(3x - 9) = 0$$

$$2x = 0, \qquad 2x + 1 = 0, \qquad x - 2 = 0, \qquad 3x - 9 = 0$$

$$x = 0 \qquad 2x = -1 \qquad x = 2 \qquad 3x = 9$$

$$x = -\tfrac{1}{2} \qquad\qquad x = 3$$

There are four solutions:

$$x = 0, \qquad x = -\tfrac{1}{2}, \qquad x = 2, \qquad x = 3$$

It's your turn again.

Problem 2

Solve: $5x(3x - 1)(x + 7) = 0$.

Solution:

Answer: $x = 0, \frac{1}{3}, -7$

Here are five more examples.

EXAMPLE 6

Solve: $2x^3 - 10x^2 - 28x = 0$.

Solution: You will save yourself time and energy if you take out the **largest** common factor first.

$$2x^3 - 10x^2 - 28x = 0$$
$$2x(x^2 - 5x - 14) = 0$$
$$2x(x - 7)(x + 2) = 0$$

$2x = 0,$ $x - 7 = 0,$ $x + 2 = 0$

$x = 0$ $x = 7$ $x = -2$

EXAMPLE 7

Solve: $x^5 - x^4 - 12x^3 = 0$.

Solution: $x^5 - x^4 - 12x^3 = 0$
$$x^3(x^2 - x - 12) = 0$$
$$x^3(x - 4)(x + 3) = 0$$

$x^3 = 0,$ $x - 4 = 0,$ $x + 3 = 0$

$x = 0$ $x = 4$ $x = -3$

EXAMPLE 8

Solve: $x^3 + x^2 - x = 0$.

Solution: $x^3 + x^2 - x = 0$

$x(x^2 + x - 1) = 0$

Since $x^2 + x - 1$ cannot be readily factored, go to step 3.

$x = 0, \quad x^2 + x - 1 = 0$

This is now a second-degree equation, and the quadratic formula can be used.

Since $x^2 + x - 1 = 0$

and $ax^2 + bx + c = 0,$ then $a = 1, b = 1, c = -1.$

Substituting into the quadratic formula gives

$$x = \frac{-b \pm \sqrt{b^2 - 4ac}}{2a}$$

$$= \frac{-1 \pm \sqrt{1 - 4(-1)}}{2}$$

$$= \frac{-1 \pm \sqrt{5}}{2}$$

There are three solutions:

$$x = 0, \qquad x = \frac{-1 + \sqrt{5}}{2}, \qquad x = \frac{-1 - \sqrt{5}}{2}$$

EXAMPLE 9

Solve: $2x^4 + 2x^3 + 2x^2 = 0$.

Solution: $2x^4 + 2x^3 + 2x^2 = 0$

$2x^2(x^2 + x + 1) = 0$

$2x^2 = 0, \quad x^2 + x + 1 = 0$

$x^2 = 0 \qquad$ Use the quadratic formula with $a = 1, b = 1, c = 1.$

$x = 0 \qquad x = \dfrac{-1 \pm \sqrt{1 - 4}}{2}$

$$= \frac{-1 \pm \sqrt{-3}}{2}$$

Therefore the only solution within the real numbers to this equation is $x = 0$.

EXAMPLE 10

Solve: $2x^3 - 6x^2 + 2x = 0$.

Solution:

$$2x^3 - 6x^2 + 2x = 0$$

$$2x(x^2 - 3x + 1) = 0$$

$$2x = 0, \quad x^2 - 3x + 1 = 0$$

$x = 0$ Use the quadratic formula with $a = 1, b = -3, c = 1$.

$$x = \frac{-b \pm \sqrt{b^2 - 4ac}}{2a}$$

$$= \frac{-(-3) \pm \sqrt{9 - 4(1)}}{2}$$

$$= \frac{3 \pm \sqrt{5}}{2}$$

The three solutions are

$$x = 0, \qquad x = \frac{3 + \sqrt{5}}{2}, \qquad x = \frac{3 - \sqrt{5}}{2}$$

So far, most of the examples have had a common factor that then reduced the one factor to a second-degree expression, and the quadratic formula could be used. But what happens if there is no common factor? At this stage try factoring by grouping as defined in Unit 20; if that fails, quit.

EXAMPLE 11

Solve: $x^3 - x^2 - 4x + 4 = 0$.

Solution:

Since the left-hand side has no common factor but does have four terms, try factoring by grouping.

$$x^3 - x^2 - 4x + 4 = 0$$

$$x^2(x - 1) - 4(x - 1) = 0$$

$$(x - 1)(x^2 - 4) = 0$$

$$(x - 1)(x - 2)(x + 2) = 0$$

$$x - 1 = 0, \qquad x - 2 = 0, \qquad x + 2 = 0$$

$$x = 1 \qquad\qquad x = 2 \qquad\qquad x = -2$$

EXAMPLE 12

Solve: $x^3 + 2x^2 - x - 2 = 0$.

Solution:
$$x^3 + 2x^2 - x - 2 = 0$$
$$x^2(x + 2) - (x + 2) = 0$$
$$(x + 2)(x^2 - 1) = 0$$
$$(x + 2)(x + 1)(x - 1) = 0$$
$$x + 2 = 0, \qquad x + 1 = 0, \qquad x - 1 = 0$$
$$x = -2 \qquad\qquad x = -1 \qquad\qquad x = 1$$

EXAMPLE 13

Solve: $2x^3 - 5x^2 + 7x - 23 = 0$.

Solution: This equation cannot readily be factored—grouping does not help, nor is there a common factor. Therefore the best we can do at this time is to say there are at most three real solutions (since it is a third-degree equation) but that we have no idea what they are.

Unit 23 has attempted to illustrate the basic technique used to solve third-degree and higher degree equations by factoring. The approach allows us to solve equations that are factorable but offers no help for the many equations we are unable to factor.

EXERCISES

Solve for x:

1. $(3x - 1)(x + 2)(x - 5) = 0$

2. $x^3 + 3x^2 - 10x = 0$

3. $(x + 1)(x - 7)(x - 3) = 0$

4. $x^3 - 49x = 0$

5. $x^4 + 16x^3 + 64x^2 = 0$

6. $10x^3 - 3x^2 - x = 0$

7. $x^3 = -6x - 5x^2$

8. $3ax^3 - 24ax^2 - 99ax = 0$

9. $(x - 3)^2 + 5(x - 3) = 14$

10. $2x^3 - x^2 + 14x - 7 = 0$

11. $x^3 + 2x^2 - x - 2 = 0$

12. $4x^3 + 18x^2 + 14x = 0$

13. $\dfrac{x^2}{x-4} = \dfrac{16}{x-4}, \quad x \neq 4$

14. $x^6 - 64 = 0$

15. $x^4 - x^3 + 8x - 8 = 0$

16. $x^2 + cx - 3x - 3c = 0$

17. $x^3 - 8x^2 + 17x - 10 = 0$

18. $x^4 - 81 = 0$

19. $x^4 - 3x^2 - 54 = 0$

20. $3x^3 + 2x^2 - 12x - 8 = 0$

UNIT 24

Graphing Linear Equations in Two Variables

The purpose of this unit is to provide you with an understanding of how to graph linear equations. When you have finished the unit, you will be able to identify linear equations in two variables, distinguish them from other types of equations, and graph them.

RECTANGULAR OR CARTESIAN COORDINATE SYSTEM OF GRAPHING

We will use a rectangular or Cartesian coordinate system for all graphs. You probably remember the basic facts:

Usually the horizontal axis is labeled x.
The vertical axis is labeled y.
The origin, O, is the point where the axes cross.
Each of the four sections is called a quadrant.
Quadrants are numbered I, II, III, and IV as shown below.

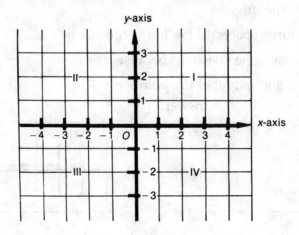

The coordinates of a point are written as an ordered pair, (x, y), where x is the number of horizontal spaces the point is from the origin:

> to the right if x is positive,
> to the left if x is negative;

and y is the number of vertical spaces the point is from the origin:

> up if y is positive,
> down if y is negative.

EXAMPLE 1

Plot: $A(2, 3)$.

Solution: 1. Start at the origin

2. Move two spaces to the right because the x-coordinate is 2.

3. Then move three spaces up because the y-coordinate is 3.

4. Draw a dot.

Answer:

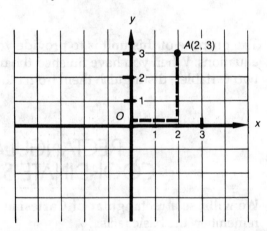

EXAMPLE 2

Plot: $B(-3, 1)$.

Solution: 1. Start at the origin.

2. Move three spaces to the left because x is -3.

3. Then move one space up because y is 1.

4. Draw a dot and label the point.

Answer:

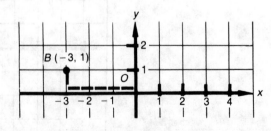

EXAMPLE 3

Plot: $C(2.5, -2)$.

Solution: 1.   Start at the origin.

2.   Move 2.5 spaces to the right because x is 2.5.

3.   Then move two spaces down because y is -2.

4.   Draw a dot and label the point.

Answer:

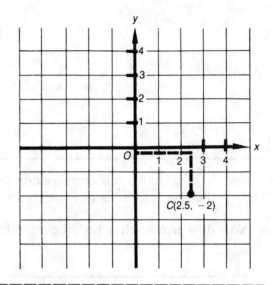

When plotting points yourself, there is no need to **draw** the dotted lines as I did in the examples. That was done only to show the procedure.

Now try to plot some points yourself. Use the graph provided.

Problem 1 $D(-4, 3)$.

Problem 2 $E(5, 2)$.

Problem 3 $F(\frac{3}{2}, -3)$.

Problem 4 $G(-2, 0)$.

Problem 5 $H(0, -5)$.

Problem 6 $J(-4.5, -2)$.

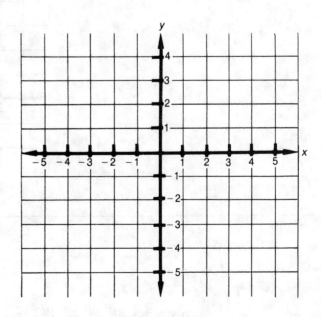

Answer:

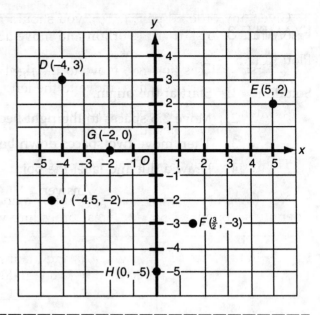

Note: Each ordered pair corresponds to one point on the graph, and for each point on the graph there is one ordered pair.

Now do a problem in which the question is reversed.

Problem 7

Find the coordinates for each of the points on the graph below:

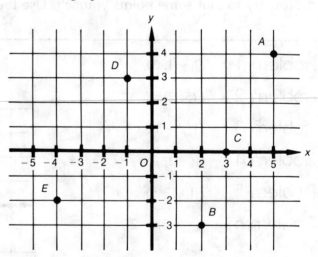

Answers: $A(5, 4)$ $B(2, -3)$ $C(3, 0)$ $D(-1, 3)$ $E(-4, -2)$

Given any ordered pair, (x, y), you should now be able to plot it. Remember that the procedure is **to start at the origin and move according to the following guide:**

First: if x is positive, move to the right;
             if x is negative, move to the left.

Then: if y is positive, move up;
            if y is negative, move down.

Then draw a dot and label the point.

Next we will consider what is meant by a linear equation in two variables, how we can identify such equations, and what procedure we can use to graph them.

> Definition: $ax + by = c$, with a, b, and c being real numbers, a and b not both zero, is
> **a linear equation in two variables.**

In other words, in a linear equation in two variables:

1.   There are two variables.

2.   Each variable is involved in only the four fundamental operations—addition, subtraction, multiplication, and division.

3.   Neither variable is raised to any power other than 1.

4.   Neither variable appears in any denominator.

5.   No term contains a product of the two variables.

Here are some examples of linear equations in two variables:

$$3x + 2y = 5$$
$$\tfrac{1}{2}y + 7x = -\sqrt{13}$$
$$x - y = 0$$
$$\frac{x}{3} = 11y - 201$$

Here are some examples that are *not* linear equations in two variables:

$$3x^2 + 2y = 7$$
$$-5x + \sqrt{y} = 10$$
$$\frac{1}{x} - 7 = y$$
$$-13xy + y = 321$$
$$-2x - 3y + z = 15$$

Before proceeding, determine why each of the above is *not* a linear equation in two variables.

In the rest of this unit, we will consider only linear equations in two variables.

You should now be able to identify a linear equation and to distinguish it from equations that are not linear. Remember that in any linear equation the two variables:

> are involved in only the four fundamental operations,
> are never multiplied together,
> are never in the denominator,
> are never raised to a power other than 1.

Now we are ready to begin graphing linear equations.

Definition: The **graph of an equation** is the set of all points whose coordinates satisfy the equation.

The graph of a linear equation is a straight line.

To graph a linear equation, locate three points whose coordinates satisfy the equation and connect them with a straight line. Actually only two points are needed; the third point serves as a check.

To locate each point:

1.   Select some convenient value for x.

2.   Substitute this value into the equation.

3.   Solve for y.

EXAMPLE 4

Graph: $x + 2y = 6$.

Solution: $x + 2y = 6$ is a linear equation; therefore its graph will be a straight line.

Locate the first point:

1.   Select a convenient value for x.   Let $x = 2$.

2.   Substitute into the equation. $2 + 2y = 6$

3.   Solve for y. $2y = 4$

                                                                      $y = 2$

Then $(2, 2)$ is a point on the graph of the equation because it satisfies the equation:

$$2 + 2(2) \overset{?}{=} 6$$

$$6 = 6$$

Repeat the procedure to find a second point:

1.   Select a value for x.   Let $x = 0$.

2.   Substitute. $0 + 2y = 6$

3.   Solve for y. $y = 3$

Thus (0, 3) is a second point on the graph because (0, 3) satisfies the equation:

$$0 + 2(3) \overset{?}{=} 6$$

$$6 = 6$$

Repeat the procedure to find a third point:

1. Select a value for x. Let $x = 4$.
2. Substitute. $4 + 2y = 6$
3. Solve for y. $2y = 2$
  $y = 1$

Then (4, 1) is a third point on the graph because (4, 1) satisfies the equation:

$$4 + 2(1) \overset{?}{=} 6$$

$$6 = 6$$

Plot the three points and connect with a straight line.

Answer:

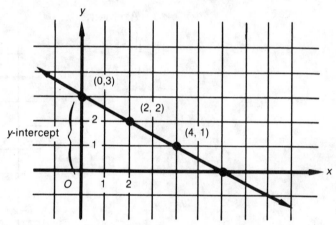

Note that there was nothing special about the three values selected for x. It often happens that $x = 0$ is a convenient value to use. The other two were selected so that y would result in an integral value.

_____

The point where the line crosses the y-axis is called the y-intercept. In other words, the y-intercept is the value of y when $x = 0$. In the above example the y-intercept is 3.

EXAMPLE 5

Graph: $3x - y = 5$.

_____

Solution: $3x - y = 5$ is a linear equation; the graph will be a straight line. Locate three points on the line.

Let $x = 0$; then $3(0) - y = 5$,

$$0 - y = 5,$$

$$-y = 5,$$

$$y = -5 \text{ and } (0, -5) \text{ is one point.}$$

Let $x = 2$; then $3(2) - y = 5$,

$$6 - y = 5,$$

$$1 = y \text{ and } (2, 1) \text{ is a second point.}$$

Let $x = -2$; then $3(-2) - y = 5$,

$$-6 - y = 5,$$

$$-y = 11,$$

$$y = -11 \text{ and } (-2, -11) \text{ is a third point.}$$

Plot the three points, and connect them with a straight line.

Answer:

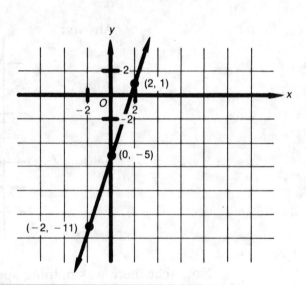

Here's a problem for you.

Problem 8

Question: What is the y-intercept of Example 5?

Solution:

Answer: -5

I'll do another example.

EXAMPLE 6

Graph: $2x + 3y = 21$.

Solution: $2x + 3y = 21$ is a linear equation; the graph will be a straight line. Locate three
points whose coordinates satisfy the equation.

Let $x = 0$; then $2(0) + 3y = 21$,

$$3y = 21,$$

$$y = 7 \text{ and } (0, 7) \text{ is one point.}$$

Let $x = 3$; then $2(3) + 3y = 21$,

$$6 + 3y = 21,$$

$$3y = 15,$$

$$y = 5 \text{ and } (3, 5) \text{ is a second point.}$$

Let $x = -2$; then $2(-2) + 3y = 21$,

$$-4 + 3y = 21,$$

$$3y = 25,$$

$$y = 8\tfrac{1}{3} \text{ and } (-2, 8\tfrac{1}{3}) \text{ is a third point.}$$

Plot the three points, and connect them with a straight line.

Answer:

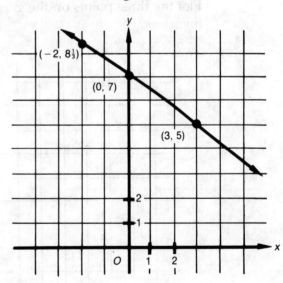

Now you try to graph a linear equation.

Problem 9

Graph: $5x + 2y = 10$.

Solution: $5x + 2y = 10$ is a linear equation; the graph will be a straight line.

Locate three points whose coordinates satisfy the equation.

Let $x = -2$; then

Let $x = 0$; then

Let $x = 2$; then

Plot the three points on the graph provided, and connect them with a straight line.

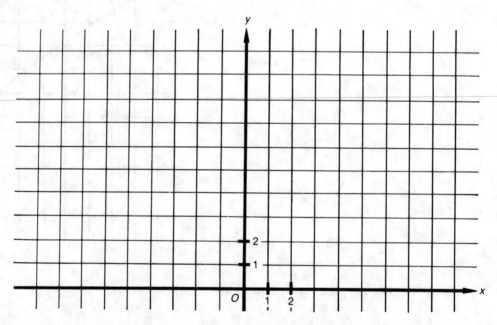

Reminder: Be sure to put the arrows at each end to indicate that the line continues on in both directions.

Answer:

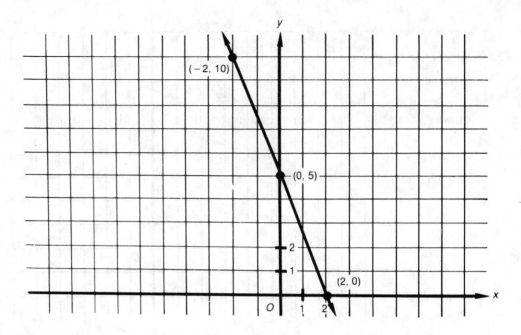

Try the next graph without any clues.

Problem 10

Graph: $-x + y = 7$. Identify the y-intercept.

Solution:

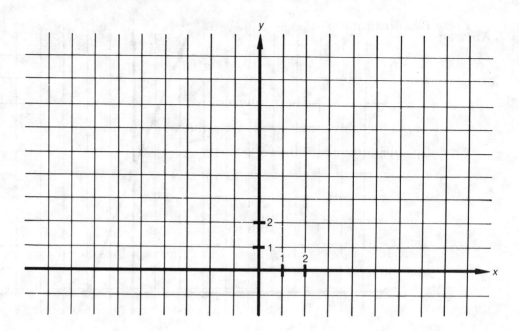

Answer:

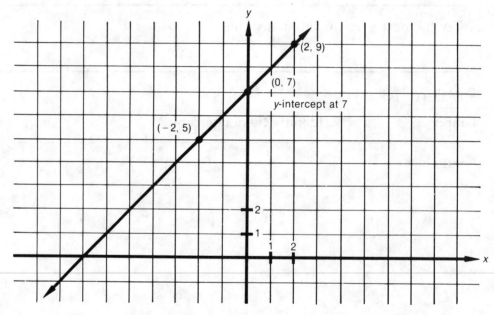

The *y*-intercept is 7.

The point where the line crosses the *x*-axis is called the *x*-intercept. In other words, the *x*-intercept is the value of *x* when $y = 0$.

EXAMPLE 7

Find the *x*- and *y*-intercepts for $5x + 2y = 10$.

Solution: The *x*-intercept is the value of *x* when $y = 0$.

By substitution: $5x + 2(0) = 10$

$$5x = 10$$

$$x = 2$$

The x-intercept is 2.

Recall that the y-intercept is the value of y when $x = 0$.

By substitution: $5(0) + 2y = 10$

$$2y = 10$$

$$y = 5$$

The y-intercept is 5.

Refer to the graph of Problem 9 to verify that the line crosses the x-axis at 2 and the y-axis at 5.

Often the x- and y-intercepts are wise choices for two of the three points used to graph a line.

Whenever possible, a linear equation should be simplified *before* locating the three points used to graph it.

EXAMPLE 8

Graph: $\left. \begin{array}{l} \dfrac{x}{3} + y = 3 \\[2mm] 4x + 2y = 0 \end{array} \right\}$ on the same set of axes.

Solution: Consider $(x/3) + y = 3$. Simplify the equation by clearing of fractions first.

$$x + 3y = 9$$

It is a linear equation, the graph will be a straight line. Locate three points. I will start with the intercepts.

The y-intercept is where $x = 0$, $0 + 3y = 9$

$$y = 3$$

Therefore the y-intercept is 3.

The x-intercept is where $y = 0$, $x + 3(0) = 9$

$$x = 9$$

Therefore the x-intercept is 9. Locate one additional point as a check.

Let $x = 3$; then $3 + 3y = 9$,

$$3y = 6,$$

$$y = 2 \text{ and the point is } (3, 2).$$

Consider the second equation: $4x + 2y = 0$.

The equation can be simplified by dividing through by 2:

$$2x + y = 0$$

It is a linear equation; the graph will be a straight line. Locate three points and graph. Three such points are $(0, 0)$, $(2, -4)$, and $(-2, 4)$. Notice that the x-intercept and y-intercept are both 0.

Answer:

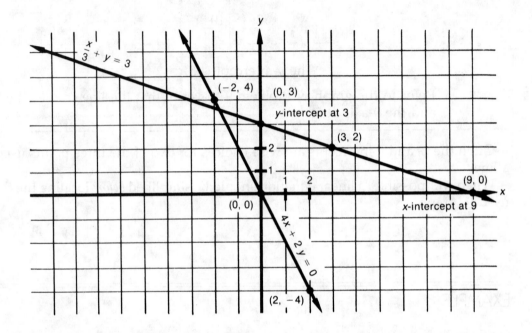

Try graphing the next group of equations yourself before looking at the answer.

Problem 11

Graph: $\left.\begin{array}{l} 7y - 7x = 35 \\[6pt] x + \dfrac{y}{2} = 4 \\[6pt] y = -\tfrac{1}{6}x + 3 \end{array}\right\}$ on the same set of axes.

Solution   (Use graph paper on page 177.):

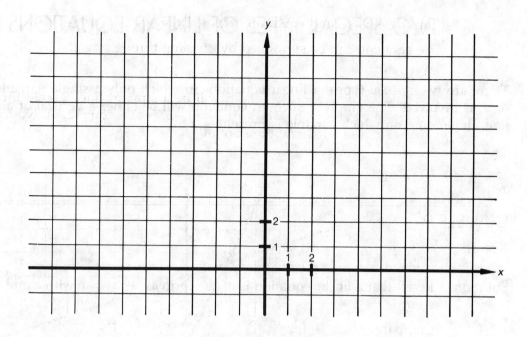

Answer:

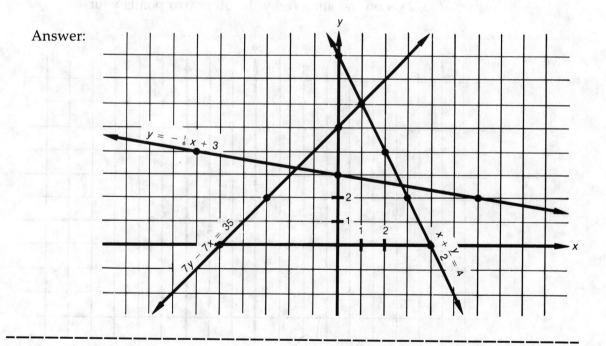

TWO SPECIAL TYPES OF LINEAR EQUATIONS

There are two special types of linear equations in which only a single variable, either x or y, is involved. The graph of such an equation will be either a vertical or a horizontal line, depending on which variable is present.

EXAMPLE 9

Graph: $x = 3$.

Solution: Recall that a linear equation is of the form $ax + by = c$ with a and b not both zero.

Compare: $\boxed{a}\, x\; \boxed{+\; b}\, y\; =\; \boxed{c}$
$x = \boxed{3}$ so $a = 1$, $b = 0$, $c = 3$.

Therefore $x = 3$ is classified as a linear equation; its graph will be a straight line. Locate three points whose coordinates satisfy the equation.

For the moment, think of the equation as $x + 0y = 3$. Three points whose coordinates satisfy the equation are $(3, 2)$, $(3, 5)$, and $(3, 0)$.

To verify $(3, 2)$: $x + 0y = 3$

$$3 + 0(2) = 3$$

$$3 = 3$$

Thus point $(3, 2)$ is on the line. Verify the other two points yourself.

Answer:

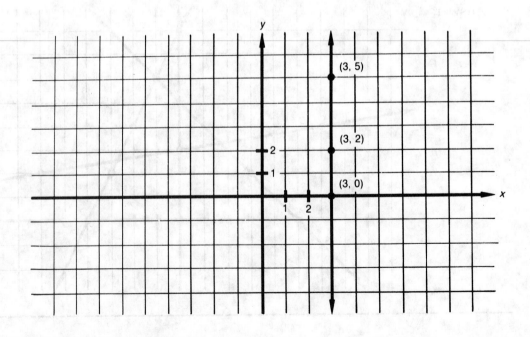

Observe: The graph of $x = 3$ is a vertical line.

Each point on the line has an x-coordinate of 3.

The x-intercept is 3.

There is no y-intercept.

> The graph of a linear equation $x = r$, where r is a real number, is a *vertical line* with x-intercept of r.

EXAMPLE 10

Graph: $x = -4$.

Solution: The graph of a linear equation $x = -4$ is a vertical line with x-intercept of -4.

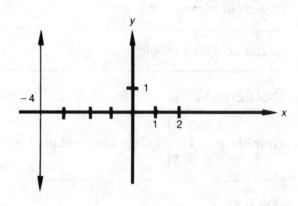

Answer:

EXAMPLE 11

Graph: $y = 2$.

Solution: By a similar line of reasoning,

compare: $a\,x + b\,y = c$

$y = 2$ so $a = 0$, $b = 1$, $c = 2$.

Thus $y = 2$ is also a linear equation; its graph will be a straight line.

If you think of the equation as $0x + y = 2$, three points whose coordinates satisfy the equation are $(5, 2)$, $(-3, 2)$, and $(0, 2)$. This time the graph is a horizontal line with each point having a y-coordinate of 2. The y-intercept is 2.

Answer:

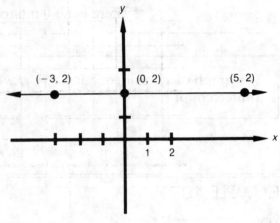

> The graph of a linear equation $y = p$, where p is a real number, is a *horizontal line* with y-intercept of p.

You do a final problem.

Problem 12

Graph: $\left.\begin{array}{l} x = 5 \\ y = 4 \\ y = -3 \end{array}\right\}$ on the same set of axes.

Solution:

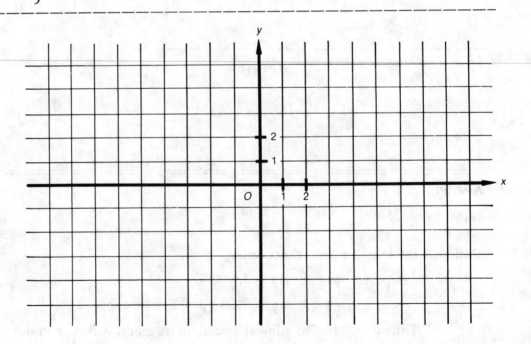

Answer:

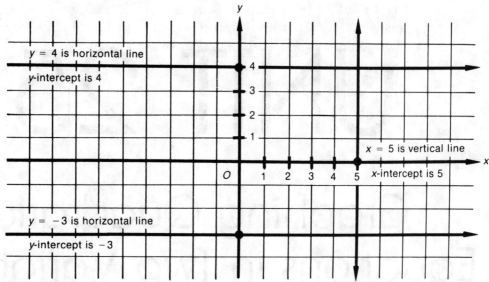

You should now be able to identify and graph linear equations in two variables by locating three points whose coordinates satisfy the equation and connecting them with a straight line. Each of the three points may be found by:

selecting some convenient value for x,
substituting this value into the equation, and
solving for y.

And you should be able to identify both the x- and y-intercepts of a linear equation.

Also, you should recognize the two special types of linear equations and their graphs:

1. The graph of $x = r$, where r is a real number, is a *vertical line* with x-intercept of r.

2. The graph of $y = p$, where p is a real number, is a *horizontal line* with y-intercept of p.

Before beginning the next unit you should graph the following equations.

EXERCISES

Identify the x-intercept, the y-intercept, and graph each equation:

1. $x + y = 8$

2. $3x - 4y = 12$

3. $7x + y = 10$

4. $y = 2$

5. $y - 2x = 0$

6. $x + 3y = 21$

7. $y = \frac{2}{7}x + 4$

8. $x = -3$

9. $-3x = 2y + 4$

10. $y = \frac{2}{5}x - 1$

If additional practice is needed:

Drooyan and Wooton, Unit 6, page 119, exercise 2
Lial and Miller, page 56, problems 1–12
Rees, Sparks, and Rees, page 106, problems 1–12
Rich, pages 120 and 121, problems 1, 5, 6

UNIT 25

Graphing Quadratic Equations in Two Variables

In this unit you will learn to recognize quadratic equations in two variables and to graph them using a minimum number of well-chosen points.

> **Definition:** $y = ax^2 + bx + c$, with a, b, and c being real numbers, $a \neq 0$, is called a **quadratic** or **second-degree equation in two variables.**

In other words, a second-degree equation in two variables must contain two variables, a squared term, x^2, and no higher powered term.

In this unit we will deal only with second-degree or quadratic equations in two variables.

Recall from Unit 24 that the graph of an equation was defined as the set of all points, (x, y) whose coordinates satisfy the equation.

There are basically two ways to graph quadratic equations:

1. Find and plot a large number of points—this can be time consuming, and the key points missed.

2. Plot a few well-chosen points based on knowledge of quadratic equations—this is the approach we will use.

Before continuing, let me illustrate the first approach, using the simplest possible second-degree equation in two variables.

EXAMPLE 1

Graph: $y = x^2$.

Solution: Find and plot a large number of points whose coordinates satisfy the equation.

Let $x = -2$; then $y = (-2)^2 = 4$, which gives point $(-2, 4)$.

Let $x = -1$; then $y = (-1)^2 = 1$, which gives point $(-1, 1)$.

Let $x = -\frac{1}{2}$; then $y = (-\frac{1}{2})^2 = \frac{1}{4}$, which gives $(-\frac{1}{2}, \frac{1}{4})$.

Let $x = 0$; then $y = (0)^2 = 0$, which gives $(0, 0)$.

Let $x = \frac{1}{2}$; then $y = (\frac{1}{2})^2 = \frac{1}{4}$, which gives $(\frac{1}{2}, \frac{1}{4})$.

Let $x = 1$; then $y = 1^2 = 1$, which gives $(1, 1)$.

Let $x = 2$; then $y = 2^2 = 4$, which gives $(2, 4)$.

Answer:

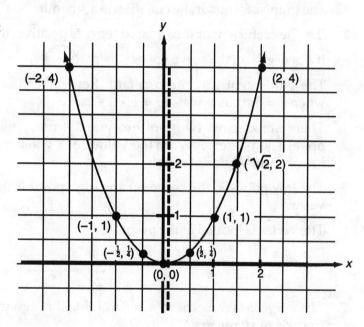

If you are not convinced that this is the graph, locate as many more points as you need to be sure. You might let x equal such values as 1.5, 3, -4, or even $\sqrt{2}$, and plot the corresponding points.

Be sure to notice that every ordered pair whose coordinates satisfy the equation, $y = x^2$, must correspond to some point on the curve. And every point on the curve must have coordinates that satisfy the equation.

As you probably observed, plotting points could become quite time consuming if the quadratic equation was more complicated, so we will move to the second method.

Definition: $y = ax^2 + bx + c$ is called the **standard form** of a quadratic equation in two variables.

Note: **All** terms are on the right side of the equal sign with **only** y on the left.

Fortunately, the graph of a quadratic equation is very predictable. The following are some basic facts that can be used when graphing a quadratic equation in two variables.

BASIC FACTS ABOUT THE GRAPH OF A QUADRATIC EQUATION IN STANDARD FORM: $y = ax^2 + bx + c$

1. The graph of a quadratic equation is a smooth, $\smile$-shaped curve called a parabola.

2. If a, the coefficient of the squared term, is positive, the curve opens up: $\smile$.

   If a is negative, the curve opens down: $\frown$.

3. The y-intercept is c, the constant. Remember that the y-intercept is the value of y when $x = 0$; thus $y = a \cdot 0^2 + b \cdot 0 + c = c$.

4. There are **at most two x-intercepts**, "at most" meaning there can be two, one, or none. The x-intercepts are the values of x when $y = 0$; thus the solution to $0 = ax^2 + bx + c$ yields the x-intercepts.

5. The low point on the curve (or the high point if the curve opens down) is called the **vertex.**

6. The vertex is located at the point

$$\left(\frac{-b}{2a}, \frac{4ac - b^2}{4a} \right).$$

7. The curve is symmetric to a vertical line through the vertex. The vertical line is called the **axis of symmetry.**

On the basis of the above information about quadratic equations in two variables, it is now possible to graph such equations using only a few well-chosen points. By "few" I mean at least three, but no more than seven.

SUGGESTED APPROACH FOR GRAPHING A QUADRATIC EQUATION IN TWO VARIABLES

1. Write the quadratic equation in standard form.

2. Determine whether the parabola opens up or down.

3. Find the y-intercept at c.

4. Solve $0 = ax^2 + bx + c$ to find the x intercepts, if the equation is easily factored.

5. Find the coordinates of the vertex:

$$\left(\frac{-b}{2a}, \frac{4ac - b^2}{4a}\right).$$

6. Locate one point on either side of the vertex.

7. Plot the above points, and connect them with a smooth, ⌣-shaped curve.

Examples 2–4 illustrate this approach.

EXAMPLE 2

Graph: $y = x^2 - 5x + 4$.

Solution: 1. Write in standard form and compare:

$$y = \bigcirc x^2 \bigcirc -5 \bigcirc x \bigcirc +4 \bigcirc$$
$$y = \bigcirc a \bigcirc x^2 \bigcirc +b \bigcirc x \bigcirc +c \bigcirc$$

Thus $a = 1, b = -5, c = 4$.

Graph will be a parabola.

2. Since a is positive, curve opens up: ⌣.

3. The y-intercept is $c = 4$.

4. Find the x-intercepts by solving:

$$0 = x^2 - 5x + 4$$
$$0 = (x - 4)(x - 1)$$
$$x = 4, \qquad x = 1$$

The x-intercepts are 4 and 1.

5. Find the coordinates of the vertex:

$$\frac{-b}{2a} = \frac{-(-5)}{2(1)} = \frac{5}{2}$$

$$\frac{4ac - b^2}{4a} = \frac{4(1)(4) - (-5)^2}{4(1)} = \frac{16 - 25}{4} = \frac{-9}{4}$$

The vertex is located at $\left(\frac{5}{2}, \frac{-9}{4}\right)$

6. Locate one point on either side of the vertex. Since the vertex is at $x = \frac{5}{2} = 2.5$:

let $x = 2$; then $y = (2)^2 - 5(2) + 4 = 4 - 10 + 4 = -2$

let $x = 3$; then $y = (3)^2 - 5(3) + 4 = 9 - 15 + 4 = -2$

7. Plot the above points, and connect them with a smooth curve.

Answer:

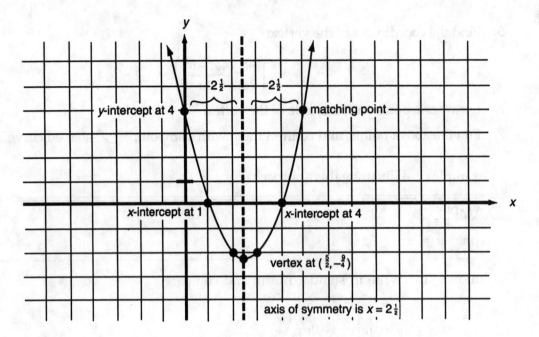

Before going on to another example, let's mention a few points about the above graph. The axis of symmetry is indicated with a dashed line. Because the curve is symmetrical about this line, there must be a point $2\frac{1}{2}$ spaces away that corresponds to the y-intercept at 4. The actual coordinates of the point are not important—only its location, which is shown, need concern us.

The arrows on each end of the parabola indicate that the curve continues upward.

_____

EXAMPLE 3

Graph: $y = -x^2 + 2x + 3$.

_____

Solution: 1.   Write in standard form and compare:

$$y = -x^2 + 2x + 3$$
$$y = ax^2 + bx + c$$

Thus $a = -1$, $b = 2$, $c = 3$.

Graph will be a parabolic curve.

2.   Since a is negative, curve will open down: ⌢ ↘.

3.   The y-intercept is $c = 3$.

4. Find the x-intercepts, at most two, by solving:

$$0 = -x^2 + 2x + 3$$

$$0 = x^2 - 2x - 3$$

$$0 = (x - 3)(x + 1)$$

$$x = 3, \qquad x = -1$$

The x-intercepts are 3 and -1.

5. Find the coordinates of the vertex:

$$\frac{-b}{2a} = \frac{-(2)}{2(-1)} = \frac{-2}{-2} = 1$$

$$\frac{4ac - b^2}{4a} = \frac{4(-1)(3) - (2)^2}{4(-1)} = \frac{-12 - 4}{-4} = \frac{-16}{-4} = 4$$

The vertex is located at (1, 4).

6. Locate one point on either side of the vertex.
    In this example step 6 may be omitted because we already have points on either side of the vertex, the y-intercept, its matching point, and the two x-intercepts.

7. Plot the above points, and connect them with a smooth ⌣-shaped curve.

Answer:

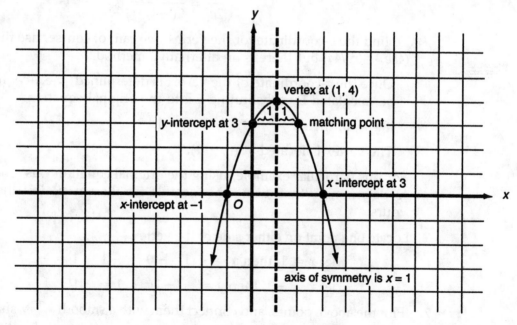

I'll do one more example before asking you to try to graph one.

EXAMPLE 4

Graph: $y = -x^2 - 1$.

Solution: 1. Write in standard form and compare:

$$y = (-)x^2 \quad (0) \quad (-1)$$
$$y = (a)x^2 + (b)x + (c)$$

Thus $a = -1$, $b = 0$, $c = -1$.

Graph will be a parabolic curve.

2. Since a is negative, curve will open down: ⌒.

3. The y-intercept is $c = -1$.

4. Find the x-intercepts, at most two, by solving:

$$0 = -x^2 - 1$$

$$x^2 = -1 \text{ has no solution}$$

Therefore there are no x-intercepts.

5. Find the coordinates of the vertex:

$$\frac{-b}{2a} = \frac{0}{2(-1)} = 0$$

To find the y-coordinate for the vertex, we can, of course, use the formula: $(4ac - b^2)/4a$, but there is an alternative method.

Once the x-coordinate of the vertex is determined, the y-coordinate can be calculated by substituting x into the original equation:

if $x = 0$, then $y = -(0)^2 - 1 = -1$.

The vertex is located at $(0, -1)$.

For this particular example, using the original equation was an easier way to determine y than using the formula for the vertex, but the choice is yours.

6. Locate one point on either side of the vertex.

$$\text{let } x = 1; \text{ then } y = -(1)^2 - 1 = -1 - 1 = -2$$
$$\text{let } x = -1; \text{ then } y = -(-1)^2 - 1 = -1 - 1 = -2$$

7. Plot the above points, and connect them with a smooth ⌒-shaped curve.

Answer:

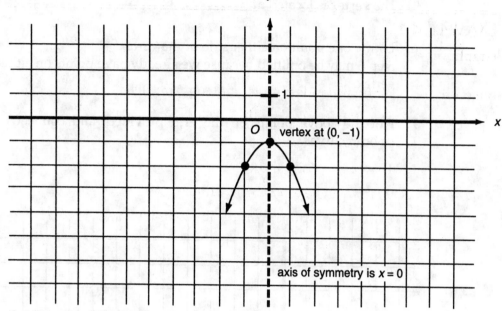

If more detail is needed, plot additional points.

Now it is time for you to try a few problems.

Problem 1

Graph: $y = -x^2 + 4x + 5$.

Solution: 1.   Write in standard form and compare:

2.   Curve will open _____ .

3.   The y-intercept is _____ .

4.   The x-intercepts, at most two, are _____ and _____ .

5. The vertex is located at _____.

6. Locate one point on either side of the vertex.
   Step 6 may be omitted because we already have points on either side of the vertex.

7. Plot the above points on the grid provided, and connect them with a smooth curve.

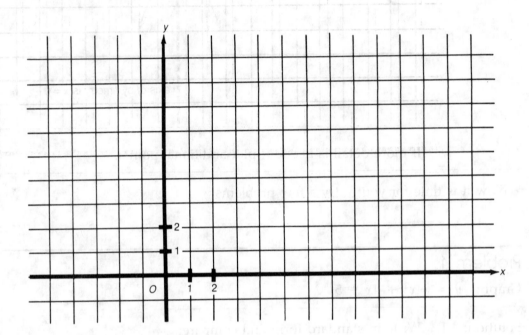

Answer:

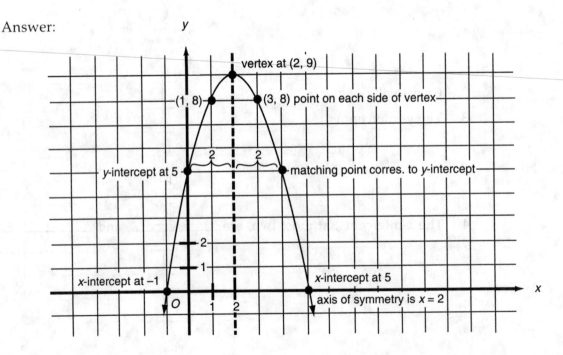

Problem 2

Graph: $y = x^2 + 2$.

Solution:

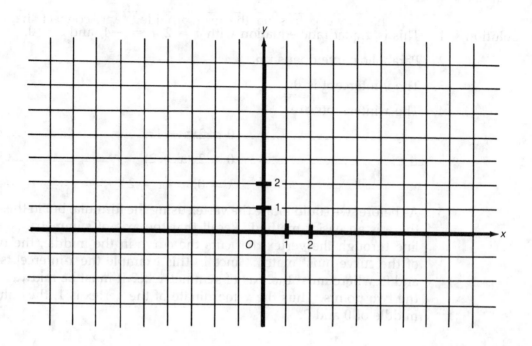

Answer:

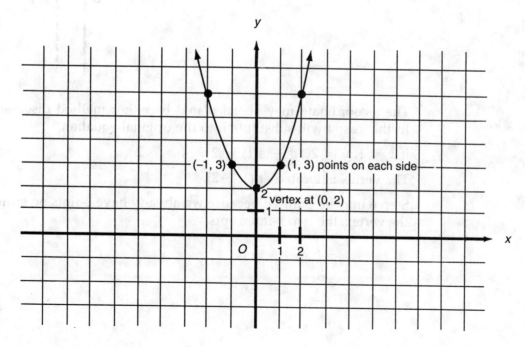

The next example illustrates still another method for locating the vertex of the parabola.

EXAMPLE 5

Graph: $y = 2x^2 - 4x$.

Solution: 1.   This is a quadratic equation with $a = 2$, $b = -4$, and $c = 0$.

2.   Parabolic curve opens up.

3.   The y-intercept is 0.

4.   The x-intercepts are:

$$0 = 2x^2 - 4x$$
$$0 = 2x(x - 2)$$
$$x = 0, \qquad x = 2$$

5.   As before, we could locate the vertex using the formula, but in this example there is an easier method. Recall that the axis of symmetry is a vertical line through the vertex that cuts the curve in the middle; the two sides of the curve must match. Since in this example the x-intercepts are at 0 and 2, where must the axis of symmetry be? It must be midway between the two points. Thus the x-coordinate of the vertex is 1, the value in the middle of 0 and 2.

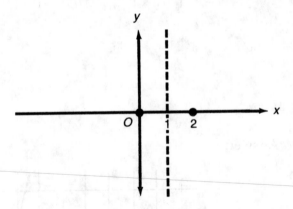

The y-coordinate may be calculated by either method discussed earlier. In this case I will substitute into the original equation:

if $x = 1$, $y = 2(1)^2 - 4(1) = 2 - 4 = -2$.

The vertex is located at $(1, -2)$.

Step 6 may be omitted because we already have points on either side of the vertex, the two x-intercepts.

Answer:

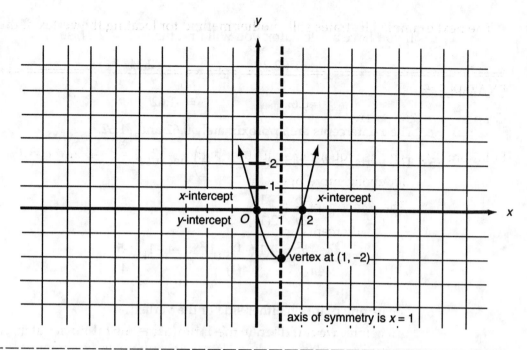

As you probably suspected, all quadratic equations are not as easy to work with as the preceding ones. Consider the next example.

EXAMPLE 6

Graph: $y = x^2 + x - 1$.

Solution: 1. This is a quadratic equation with $a = 1$, $b = 1$, and $c = -1$.

2. Parabolic curve opens up.

3. The y-intercept is -1.

4. Find the x-intercepts, at most two, by solving:

$$0 = x^2 + x - 1$$

Since the right-hand side is not easily factored, we will go on to Step 5.

Note: If the x-intercepts were required, the quadratic formula could be used to find them since we are unable to factor.

$$x = \frac{-b \pm \sqrt{b^2 - 4ac}}{2a}$$

$$= \frac{-1 \pm \sqrt{(1)^2 - 4(1)(-1)}}{2(1)}$$

$$= \frac{-1 \pm \sqrt{5}}{2} \text{ are the } x\text{-intercepts}$$

If you have a calculator, you will find that $\sqrt{5} \approx 2.24$, so

$$x = \frac{-1 + 2.24}{2} \quad \text{and} \quad x = \frac{-1 - 2.24}{2}$$
$$= 0.62 \qquad\qquad = -1.62$$

The x-intercepts are approximately 0.62 and –1.62.

5. The vertex is located at $\left(\dfrac{-1}{2}, \dfrac{-5}{4} \right)$ because:

$$\frac{-b}{2a} = \frac{-1}{2(1)} = \frac{-1}{2}$$

$$\frac{4ac - b^2}{4a} = \frac{4(1)(-1) - (1)^2}{4(1)} = \frac{-4 - 1}{4} = \frac{-5}{4}$$

6. Locate one point on either side of the vertex:
let $x = -1$; then $y = (-1)^2 + (-1) - 1 = -1$
let $x = 0$; then $y = -1$

Answer:

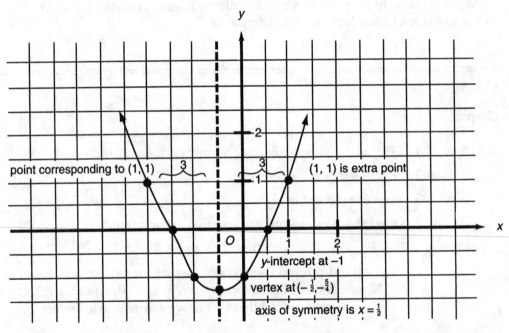

point corresponding to (1, 1)   (1, 1) is extra point
y-intercept at –1
vertex at $\left(-\frac{1}{2}, -\frac{5}{4}\right)$
axis of symmetry is $x = \frac{1}{2}$

Comments: I approximated the x-intercepts. To get a better sense of where the curve actually crossed the x-axis, I plotted a convenient extra point, (1, 1), in quadrant I. There would be a corresponding point in quadrant II. Then, by connecting the five points with a smooth curve, the location of the x-intercepts is well determined.

––

To summarize, in Unit 25 we defined (and you should now be able to identify) a second-degree equation in two variables, of the type $y = ax^2 + bx + c$ with $a \neq 0$. Also, you should be able to graph such an equation, using a minimum number of well-chosen points, based on your knowledge of quadratic equations.

Recall that, in short, the graphing approach is as follows:

1. Write the equation in standard form and compare.
2. Determine whether the parabola opens up or down.
3. Find the y-intercept at c.
4. Solve $0 = ax^2 + bx + c$ to find the x-intercepts, if the equation is easily factored.
5. Find the coordinates of the vertex:

$$\left(\frac{-b}{2a}, \frac{4ac - b^2}{4a} \right).$$

6. Locate a point on either side of the vertex, if needed.
7. Plot the points, and connect them with a smooth, ⌣-shaped curve.

Now try the following exercises.

EXERCISES

Graph each of the following by using your knowledge of quadratic equations in two variables to plot a few well-chosen points.

1. $y = x^2 - 2x - 8$
2. $y = x^2 + 2x - 8$
3. $y = -4x^2$
4. $y = -x^2 - 4x - 3$
5. $y = x^2 - 6x + 5$
6. $y = 4 + 2x^2$
7. $y = x^2 - 4x + 4$
8. $y = 25 - x^2$
9. $y = x^2 + 6x + 9$
10. $y = 2x^2 + 4x - 1$
11. $y = 5x^2 - 20x + 11$
12. $y = -x^2 + 10x$
13. $y = 3x^2 - 3x + 2$
14. $y = 2x^2 - 12x + 3$
15. $y = \frac{-1}{4}x^2 + 3x - 8$

If additional practice is needed:
Drooyan and Wooton, Unit 8, page 84, exercise 8
Lial and Miller, page 79, problems 1–18
Peters, page 268, problems 1–4
Rees, Sparks, and Rees, pages 106, problems 13–20

UNIT 26

Solving Systems of Equations

The purpose of this unit is to provide you with an understanding of systems of equations. When you have finished the unit, you will be able to solve systems of two linear equations in two variables.

Recall the following definition from Unit 24:

> Definition: $ax + by = c$, with a, b, and c being real numbers, a and b not both zero, is **a linear equation in two variables**.

By a "solution to an equation in two variables" we mean the pairs of values of x and y that satisfy the equation. The procedure outlined earlier for locating points on the graph of an equation also yields solutions to the equation. In fact, solutions are often written as ordered pairs.

There are infinitely many solutions to an equation in two variables.

EXAMPLE 1

Solve: $x + y = 8$.

- -

Solution: Recall how to locate a point:

1. Select a convenient value for x.   Let $x = 2$.

2. Substitute into the equation. $(2) + y = 8$

3. Solve for y. $y = 6$

196

Then (2, 6) is a point on the graph of $x + y = 8$

and

$x = 2$ and $y = 6$ is called a solution to the equation $x + y = 8$.

Without going through the calculations, (3, 5) is a point on the graph of $x + y = 8$

and

$x = 3$ and $y = 5$ is another solution to the equation.

There are infinitely many points on the graph

and

there are infinitely many solutions to the equation, some of which are

$$x = 1 \quad \text{and} \quad y = 7 \quad \text{or simply (1, 7)}$$
$$x = 0 \quad \text{and} \quad y = 8 \quad \text{or} \quad (0, 8)$$
$$x = 2.5 \quad \text{and} \quad y = 5.5 \quad \text{or} \quad (2.5, 5.5)$$
$$x = -3 \quad \text{and} \quad y = 11 \quad \text{or} \quad (-3, 11)$$

Find at least five more solutions yourself.

A **system of equations** means that there is more than one equation.

> Definition: The **solutions** to a system of equations are the pairs of values of x and y that satisfy *all* the equations in the system.

In this unit we will deal only with linear equations in two variables, saving quadratic equations for the next unit.

In general, the number of solutions to any system of *linear* equations is either one, none, or infinitely many.

$$\begin{cases} 2x + y = 24 \\ x - y = 6 \end{cases}$$

is an example of a system of linear equations. The solution to this system is $x = 10$ and $y = 4$ or simply (10, 4); because this pair of values satisfies *both* equations in the system.

To verify, check by substitution.

$$2x + y = 24 \qquad x - y = 6$$
$$2(10) + 4 \overset{?}{=} 24 \qquad 10 - 4 \overset{?}{=} 6$$
$$24 = 24 \qquad\qquad 6 = 6$$

The rest of the unit deals with finding such solutions.

There are numerous ways to solve systems of equations in two variables. The one developed in this unit is called **elimination by addition**. A variation of it, called **elimination by substitution**, will be explained toward the end of the unit.

ELIMINATION BY ADDITION

The procedure I will use involves four steps:

1. **Multiply (if necessary)** the equations by constants so that the coefficients of the x or the y variable are the negatives of one another.

2. **Add** the equations from step 1.

3. **Solve** the equation from step 2.

4. **Substitute** the answer from step 3 back into one of the original equations, and solve for the second variable.

Note: It is advisable to simplify and write equations in standard form before starting.

Here are some examples that illustrate the use of these steps in solving a system of linear equations.

EXAMPLE 2

Solve: $\begin{cases} x + y = 7 \\ 5x - 3y = 11 \end{cases}.$

Solution: Multiply the first equation by 3 so that the *y*-coefficients are the negatives of one another.

1. **Multiply.** $3(x + y = 7)$
   $$5x - 3y = 11$$

2. **Add.** $3x + 3y = 21$
   $$5x - 3y = 11$$

3. **Solve.** $8x\quad = 32$
   $$x\quad = 4$$

4. **Substitute back** into one of the original equations. I try to select the simpler.

   If $x = 4$ and $x + y = 7$,

   $$(4) + y = 7,$$

   $$y = 3.$$

   Answer: Solution to system is $x = 4$ and $y = 3$
   or, written another way, $(4, 3)$.

Note: Back in step 1 we could just as well have multiplied the equation by -5 so that the *x*-coefficients would have been the negatives of one another.

— —

EXAMPLE 3

Solve: $\begin{cases} 3x + 2y = 12 \\ y = 2x - 1 \end{cases}$.

— —

Solution: Write the equations in standard form.

$$3x + 2y = 12$$

$$-2x + y = -1$$

Multiply the second equation by -2 so that the *y*-coefficients are the negatives of one another.

1. **Multiply.** $3x + 2y = 12$
   $$-2(-2x + y = -1)$$

2. **Add.** $3x + 2y = 12$
   $$4x - 2y = 2$$

3. **Solve.** $7x\quad = 14$
   $$x\quad = 2$$

4.   **Substitute back** into an original equation.

If $x = 2$ and $y = 2x - 1$,

$$y = 2(2) - 1,$$

$$y = 3.$$

Answer: Solution to system is $x = 2$ and $y = 3$.

EXAMPLE 4

Solve: $\begin{cases} 3x - y = -7 \\ 5y + 5 = -5x \end{cases}.$

Solution: Simplify the second equation, and write it in standard form before starting the procedure.

Rewrite. $5x + 5y = -5$

Divide by 5. $x + y = -1$

The equations are now: $3x - y = -7$

$x + y = -1$

1.   **Multiply**—not necessary.

2.   **Add.** $3x - y = -7$

$\underline{x + y = -1}$

3.   **Solve.** $4x = -8$

$x = -2$

4.   **Substitute back** into an original equation.

If $x = -2$ and $x + y = -1$,

$$(-2) + y = -1,$$

$$y = 1.$$

Answer: Solution to system is $x = -2$ and $y = 1$.

From an algebraic point of view,

$x = -2$ and $y = 1$ is the solution to the system of equations $3x - y = -7$ and $x + y = -1$.

From a geometric point of view,

$(-2, 1)$ is the point of intersection for two lines whose equations are given by $3x - y = -7$ and $x + y = -1$.

To verify that $(-2, 1)$ is the point of intersection, graph the two lines.

Problem 1

Graph: $\begin{cases} 3x - y = -7 \\ 5y + 5 = -5x \end{cases}$.

Solution:

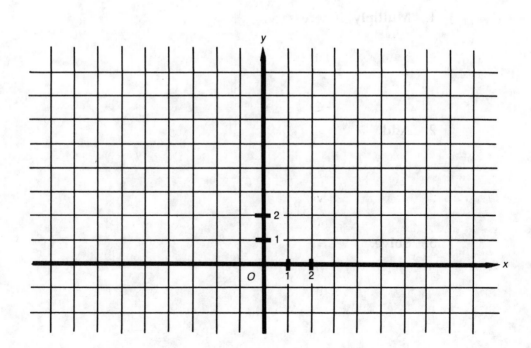

Answer:

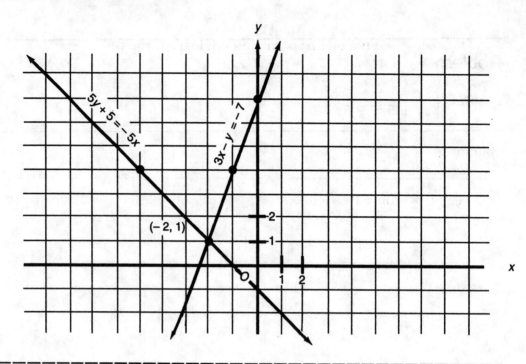

Now it is your turn to solve a system of equations.

Problem 2

Solve: $\begin{cases} 4x - y - 1 = 0 \\ 2x = 17 - y \end{cases}$.

Solution   (Hint: First write equations in standard form.):

   1. **Multiply,** if necessary.

   2. **Add.**

   3. **Solve.**

4. **Substitute back** into an original equation.

Answer: Solution to system is $x = 3$ and $y = 11$.

Verify, by doing Problem 3, that your answer to Problem 2 is the intersection of the two lines.

Problem 3

Graph: $\begin{cases} 4x - y - 1 = 0 \\ 2x = 17 - y \end{cases}$.

Solution:

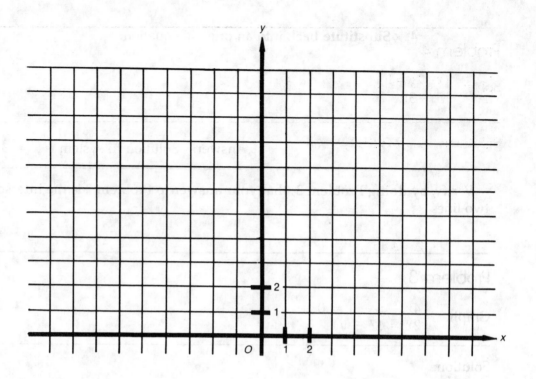

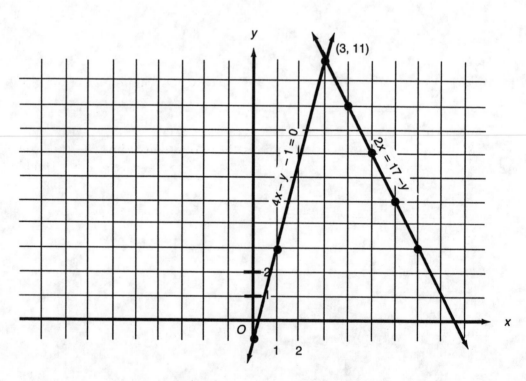

Answer:

Problem 4

Solve: $\begin{cases} x + 2y = 1 \\ 5x + 3y = 26 \end{cases}$.

Solution: 1. **Multiply.**

2. **Add.**

3. **Solve.**

4. **Substitute back** into an original equation.

Answer: Solution to system is $x = 7$ and $y = -3$.

Many times it is necessary to multiply the two equations by different constants in order to make the coefficients of the x or the y variable be the negatives of one another.

EXAMPLE 5

Solve: $\begin{cases} 2x + 3y = 9 \\ 5x + 2y = 17 \end{cases}$.

Solution: 1. **Multiply.** $-5(2x + 3y = 9)$
                                    $2(5x + 2y = 17)$

            2. **Add.** $-10x - 15y = -45$
                                    $\underline{10x + 4y = 34}$

3.   **Solve.**
$$-11y = -11$$
$$y = 1$$

4.   **Substitute back** into an original equation.

If $y = 1$ and $2x + 3y = 9$,
$$2x + 3(1) = 9,$$
$$2x = 6,$$
$$x = 3.$$

Answer: Solution to system is $x = 3$ and $y = 1$.

Did you notice that there were several possibilities for the constants used in step 1? We could just as well have multiplied the first equation by 2 and the second equation by -3. Verify, by using 2 and -3, that the solution to the system remains the same by solving Problem 5 as directed.

What other combinations could have been used?

Problem 5

Solve: $\begin{cases} 2x + 3y = 9 \\ 5x + 2y = 17 \end{cases}$.

Solution: 1.   Multiply.
$$2(2x + 3y = 9)$$
$$-3(5x + 2y = 17)$$

You finish the problem from here.

Answer: $x = 3$ and $y = 1$

Problem 6

Solve: $\begin{cases} 3x - 2y = 5 \\ -4x + 3y = 1 \end{cases}$.

Solution:

Answer: $x = 17$ and $y = 23$

As you probably suspected, all systems do not have nice integral values as solutions. Be prepared for fractions.

Problem 7

Solve: $\begin{cases} \dfrac{x}{3} + y = 3 \\ 4x + 2y = 0 \end{cases}$.

Solution (There is additional space on page 208.):

$$\text{Answer: } x = \frac{-9}{5} \text{ and } y = \frac{18}{5}$$

Recall that at the beginning of this unit it was stated that the number of solutions to a system of linear equations will be one, none, or infinitely many. Thus far, we have seen only systems with one solution. Examples 6 and 7 will examine the other two situations.

EXAMPLE 6

Solve: $\begin{cases} 3x + 4y = 2 \\ 2y = 4 - \frac{3}{2}x \end{cases}$.

Solution: First simplify the second equation, and write it in standard form.

$$2(2y = 4 - \tfrac{3}{2}x)$$
$$4y = 8 - 3x$$
$$3x + 4y = 8$$

1. **Multiply.** $-1(3x + 4y = 2)$
$$3x + 4y = 8$$

2. **Add.**
$$\begin{array}{r} -3x - 4y = -2 \\ 3x + 4y = 8 \\ \hline 0 = 6 \end{array}$$

But $0 = 6$ is a false statement; and because $0 = 6$ is a false statement, there are no values of x and y that would ever satisfy this equation.

Answer: There is no solution to the system.

The equations are said to be **inconsistent** when there is no solution to the system.

Geometrically, no solution to a system of two linear equations means that there is no point of intersection for the two lines. The lines are parallel.

Verify by graphing that the system in Example 6 represents two parallel lines.

Problem 8

Graph: $\begin{cases} 3x + 4y = 2 \\ 2y = 4 - \frac{3}{2}x \end{cases}$.

- -

Solution:

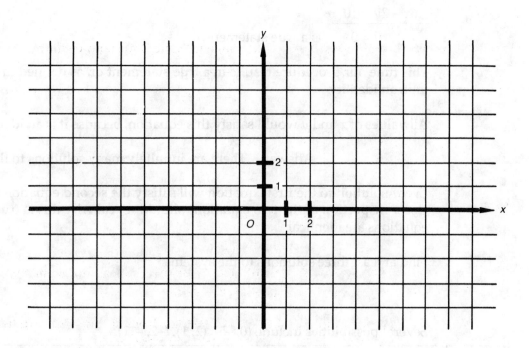

Answer:

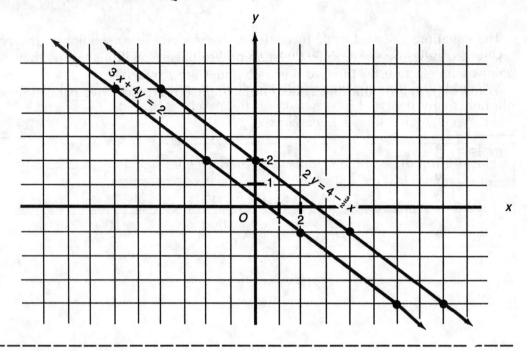

EXAMPLE 7

Solve: $\begin{cases} 3x - y = 5 \\ 6x - 2y - 10 = 0 \end{cases}$

Solution: $-2(3x - y = 5)$

$6x - 2y = 10$

$-6x + 2y = -10$

$6x - 2y = 10$

$\overline{}$

$0 = 0$ is a true statement.

This time our procedure results in a true statement or, as defined in an earlier unit, an identity.

All values of x and y would satisfy this equation, because it is an identity.

Answer: There are infinitely many solutions to the system.

Every solution to the first equation will satisfy the second equation as well. A closer inspection of the two equations should reveal why this is true; the two equations are identical.

Find at least three solutions to the system.

Several possibilities include (0, –5), (2, 1), (–2, –11).

The equations are said to be **dependent** when there are infinitely many solutions to the system.

Geometrically, "many solutions to a system of two linear equations" means there are infinitely many points of intersection for the two lines. The lines coincide.

Let's review what has just been stated.

No solution to a system is indicated when the elimination method leads to a false statement, such as $0 = 6$. The equations are said to be **inconsistent**.

Infinitely many solutions to a system are indicated when the elimination method leads to an identity, such as $0 = 0$. The equations are said to be **dependent**.

Problem 9

Solve: $\begin{cases} 5x - y = 3 \\ 2y = 10x - 6 \end{cases}$.

Solution:

Answer: There are infinitely many solutions.

Problem 10

Solve: $\begin{cases} 3x + 4y = 11 \\ 2y = 4 - \frac{3}{2}x \end{cases}$.

Solution:

Answer: There is no solution to the system.

ELIMINATION BY SUBSTITUTION

Elimination by substitution is a variation of the method we have used thus far to solve systems of equations. If one of the variables has a coefficient of 1, it is sometimes more efficient to eliminate that variable by substitution rather than by multiplication and addition. Let me illustrate what I mean with the next example.

EXAMPLE 8

Solve: $\begin{cases} y = 2x - 8 \\ 3x + 2y = 12 \end{cases}$.

Solution: Use $y = 2x - 8$ to substitute into the second equation.

$$3x + \quad 2y \qquad = 12$$
$$3x + 2(2x - 8) = 12 \qquad \text{Replace } y \text{ with } 2x - 8.$$
$$3x + \quad 4x - 16 = 12$$
$$7x \qquad\quad = 28$$
$$x = 4$$

Finish the problem as before by substituting back into an original equation to find y:

$$y = 0$$

Answer: Solution to system is $x = 4$ and $y = 0$.

I use elimination by substitution whenever possible, because it often requires less rewriting of the equations.

EXAMPLE 9

Solve: $\begin{cases} y = 3x + 1 \\ 3x + 4y = -26 \end{cases}$.

Solution: Use $y = 3x + 1$ to substitute into the second equation.

$$3x + 4y \qquad\quad = -26$$
$$3x + 4(3x + 1) = -26 \qquad \text{Replace } y \text{ with } 3x + 1.$$
$$3x + 12x + 4 = -26$$
$$15x \qquad\quad = -30$$
$$x \qquad = -2$$

and, by substituting back into an original equation,

$$y \qquad = -5$$

Answer: Solution to system is $x = -2$ and $y = -5$.

Let's do one more example, and then you can try some problems.

EXAMPLE 10

Solve: $\begin{cases} 5x - 2y = 11 \\ y = \frac{5}{2}x - 3 \end{cases}$.

Solution: Use $y = \frac{5}{2}x - 3$ to substitute into the first equation.

$$5x - 2y = 11$$

$$5x - 2(\tfrac{5}{2}x - 3) = 11 \qquad \text{Replace } y \text{ with } \tfrac{5}{2}x - 3.$$

$$5x - 5x + 6 = 11$$

$$6 = 11 \text{ is a false statement.}$$

Answer: There is no solution to system.

Here are two problems for you to solve. The first one is easy; the second is similar to the ones used as examples.

Problem 11

Solve: $\begin{cases} 3x + 5y = 14 \\ x = -2 \end{cases}$.

Solution:

Answer: $(-2, 4)$

Problem 12

Solve by substitution: $\begin{cases} y = 7x + 2 \\ x - 3y = -6 \end{cases}$

Solution:

Answer: (0, 2)

You should now be able to solve any system of two linear equations in two variables. Remember that elimination by addition is one method used to solve such systems; briefly stated, the four basic steps involved are:

1. **Multiply**, if necessary.
2. **Add.**
3. **Solve.**
4. **Substitute back.**

Also remember that the number of solutions to any system of linear equations is either one, none, or many.

Before beginning the next unit you should do the following exercises.

EXERCISES

Solve these systems, using either elimination by addition or elimination by substitition:

1. $\begin{cases} 5x + 2y = 22 \\ 3x - 2y = 10 \end{cases}$

2. $\begin{cases} 13x - 5y = -5 \\ x + y = 1 \end{cases}$

3. $\begin{cases} y = 7x + 2 \\ 2x = y + 8 \end{cases}$

4. $\begin{cases} 5x + 3y = 1 \\ -2x + 5y = 12 \end{cases}$

5. $\begin{cases} 4x - 3y = -10 \\ 5x + 2y = 22 \end{cases}$

6. $\begin{cases} 2x + y = 0 \\ x - y = 1 \end{cases}$

7. $\begin{cases} y = 7x + 2 \\ x - 3y = -6 \end{cases}$

8. $\begin{cases} -2x + 17y = 6 \\ 11x - 5y = -33 \end{cases}$

9. $\begin{cases} 5x + 2y = 50 \\ 4x - 3y = -52 \end{cases}$

10. $\begin{cases} 4x + y = 13 \\ x - 3y = 0 \end{cases}$

11. $\begin{cases} y = 2x + 3 \\ y + 2 = 4x + 1 \end{cases}$

12. $\begin{cases} x + 4y = 8 \\ y = -\frac{1}{4}x - 7 \end{cases}$

13. $\begin{cases} x - 3y = 2 \\ 4x - 10y = 10 \end{cases}$

14. $\begin{cases} 2x + \frac{1}{2}y = 2 \\ 6x - y = 1 \end{cases}$

15. $\begin{cases} x = 7 - \frac{1}{2}y \\ 8x - y = -4 \end{cases}$

16. $\begin{cases} y = -\frac{2}{5}x + 4 \\ x = \frac{1}{3}y - 7 \end{cases}$

17. $\begin{cases} 3x - 2y = 8 \\ -6x + 4y = 10 \end{cases}$

18. $\begin{cases} 11x - y = 10 \\ 2x + y = 7 \end{cases}$

19. $\begin{cases} 2a + 3b = 10 \\ 3a + 2b = 10 \end{cases}$

20. $\begin{cases} y = -\frac{2}{7}x - 1 \\ 2x = -7(y + 1) \end{cases}$

If additional practice is needed:
  Drooyan and Wooton, Unit 6, page 128, exercise 4
  Leithold, pages 314 and 315, problems 1–22
  Lial and Miller, page 163, problems 1–8
  Peters, page 204, problems 1–20
  Rees, Sparks, and Rees, pages 267, problems 1–12

UNIT 27

Solving Systems of Equations (Continued)

In Unit 26 you learned to solve systems of linear equations. When you have finished this unit, you will be able to solve systems of equations containing one linear equation and one quadratic equation.

First, recall how we define a solution:

Definition:	The **solutions** to a system of equations are the pairs of values of x and y that satisfy *all* the equations in the system.

As was previously stated, either elimination by addition or elimination by substitution may be used to solve systems of equations. Often elimination by substitution is more efficient when one of the equations is a quadratic.

ELIMINATION BY SUBSTITUTION

When one equation of the system is a quadratic, the procedure I will use involves three steps:

1. Use the quadratic equation to **substitute** into the linear equation. If necessary, write the quadratic equation in standard form first.

2. **Solve** the equation from step 1 either by factoring or by using the quadratic formula.

3. **Substitute** the answer back into one of the original equations, and solve for the second variable.

218

The examples that follow illustrate the use of these steps in solving a system of equations containing one quadratic equation.

EXAMPLE 1

Solve: $\begin{cases} -3x + 2y = 2 \\ y = x^2 \end{cases}$.

Solution: Use $y = x^2$ to substitute into the linear equation.

$$-3x + 2y = 2$$

$$\downarrow$$

$-3x + 2(x^2) = 2$ Replace y with x^2.

$-3x + 2x^2 = 2$ Solve the new equation.

$2x^2 - 3x - 2 = 0$

$(2x + 1)(x - 2) = 0$

$2x + 1 = 0$ or $x - 2 = 0$

$x = -\frac{1}{2}$ or $x = 2$

Substitute back into an original equation.

If $x = -\frac{1}{2}$	If $x = 2$
and $y = x^2$,	and $y = x^2$,
$y = (-\frac{1}{2})^2$,	$y = (2)^2$,
$y = \frac{1}{4}$.	$y = 4$.

Answer: There are two solutions to the system:

$x = -\frac{1}{2}$ and $y = \frac{1}{4}$ $(-\frac{1}{2}, \frac{1}{4})$

and

$x = 2$ and $y = 4$ $(2, 4)$

GRAPHING THE SYSTEM WHEN ONE EQUATION IS A QUADRATIC

Recall that the graph of a linear equation is a line and the graph of a quadratic equation is a parabola.

From the algebraic point of view

the system: $\begin{cases} -3x + 2y = 2 \\ y = x^2 \end{cases}$

has two solutions: $x = -\frac{1}{2}$ and $y = \frac{1}{4}$

and

$x = 2$ and $y = 4$

From a geometric point of view

$\left(-\frac{1}{2}, \frac{1}{4}\right)$ and $(2, 4)$ are the points of intersection for the line and parabola whose equations are given by $-3x + 2y = 2$ and $y = x^2$.

To verify that $\left(-\frac{1}{2}, \frac{1}{4}\right)$ and $(2, 4)$ are the points of intersection, graph the two equations as instructed in Problem 1.

Problem 1

Graph: $\begin{cases} -3x + 2y = 2 \\ y = x^2 \end{cases}$

Solution:

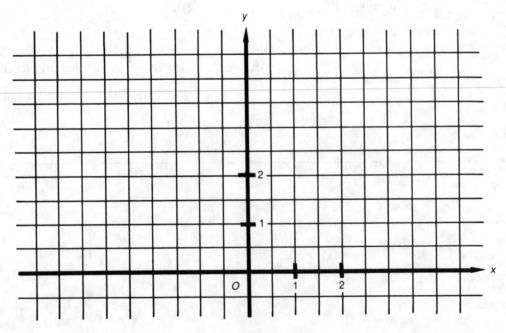

Answer:

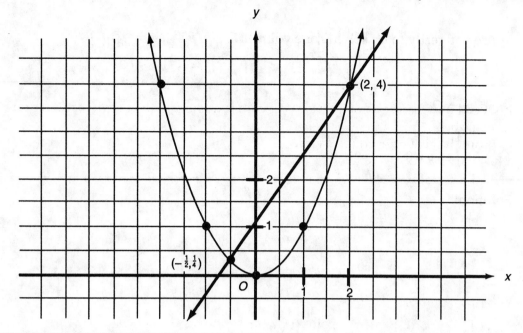

EXAMPLE 2

Solve: $\begin{cases} y = -x^2 + 2x + 3 \\ x + y = 3 \end{cases}$.

Solution: Use $y = -x^2 + 2x + 3$ to substitute into equation.

$$x \quad + \quad y \qquad = 3$$

$$x + (-x^2 + 2x + 3) = 3 \qquad \text{Solve the new equation.}$$
$$-x^2 + 3x \qquad = 0$$
$$-x(x - 3) \qquad = 0$$
$$-x = 0 \qquad \text{or} \qquad x - 3 = 0$$
$$x = 0 \qquad\qquad\qquad x = 3$$

Substitute back into an original equation.

If $x = 0$	If $x = 3$
and $x + y = 3$,	and $x + y = 3$,
$(0) + y = 3$,	$(3) + y = 3$,
$y = 3$.	$y = 0$.

Answer: There are two solutions to the system:

$x = 0$ and $y = 3$ $\qquad (0, 3)$

and

$x = 3$ and $y = 0$ $\qquad (3, 0)$.

Are you ready to try solving one yourself?

Problem 2

Solve: $\begin{cases} y = x^2 + x + 1 \\ x + y = 0 \end{cases}$.

- -

Solution:

Answer: $(-1, 1)$

- -

Verify, by doing Problem 3, that your answer to Problem 2 is the intersection of the parabola and the line.

Problem 3

Graph: $\begin{cases} y = x^2 + x + 1 \\ x + y = 0 \end{cases}$.

- -

Solution:

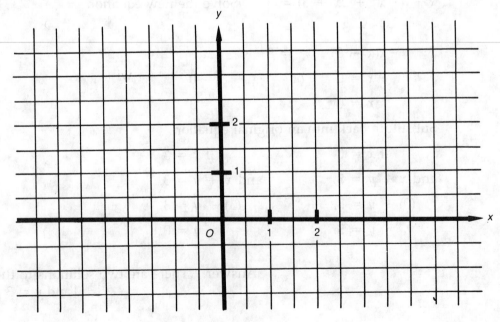

Answer:

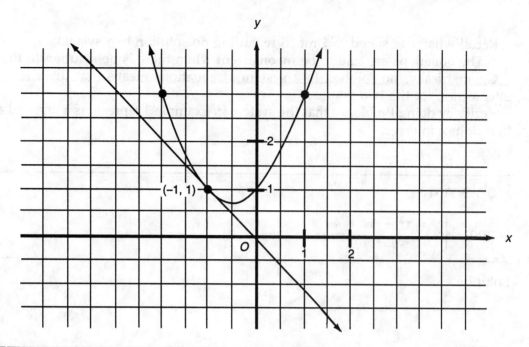

EXAMPLE 3

Solve: $\begin{cases} y = x^2 + 5x + 4 \\ 2x - 3y = 15 \end{cases}$.

Solution: Use $y = x^2 + 5x + 4$ to substitute into equation.

$$2x - \qquad 3y \qquad = 15$$

$$2x - 3(x^2 + 5x + 4) = 15$$

$$2x - 3x^2 - 15x - 12 = 15$$

$$-3x^2 - 13x - 27 = 0$$

Not readily factorable; use quadratic formula with $a = -3$, $b = -13$, and $c = -27$.

$$x = \frac{-b \pm \sqrt{b^2 - 4ac}}{2a}$$

$$= \frac{-(-13) \pm \sqrt{(-13)^2 - 4(-3)(-27)}}{2(-3)}$$

$$= \frac{13 \pm \sqrt{169 - 324}}{-6}$$

$$= \frac{13 + \sqrt{-155}}{-6}$$

No solution.

Answer: There is no solution to the system.

Recall what was stated in Unit 26 regarding no solution to a system:

The equations are said to be **inconsistent** when there is no solution to the system.

Geometrically, no solution to a system of equations means that there is no point of intersection for the two graphs.

Verify by doing Problem 4 that the system in Example 3 represents a line and a parabola that do not intersect.

Problem 4

Graph: $\begin{cases} y = x^2 + 5x + 4 \\ 2x - 3y = 15 \end{cases}$.

Solution:

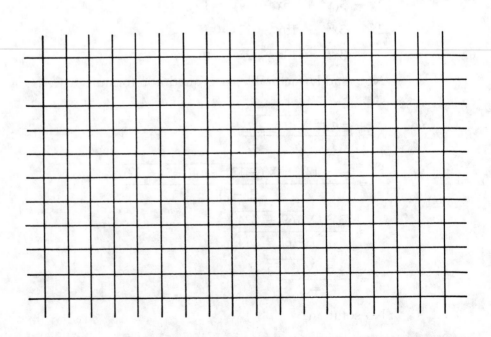

Answer:

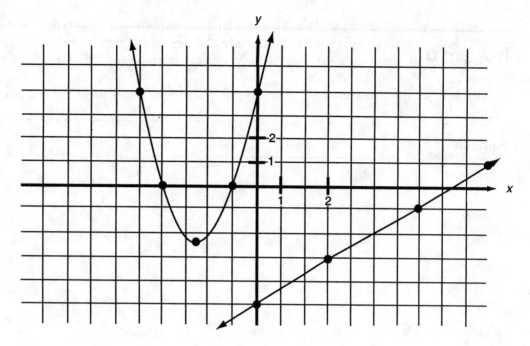

Here are a few more systems for you to solve. For additional practice, you might consider graphing them as well.

Problem 5

Solve: $\begin{cases} y = -x^2 + 4x + 5 \\ x - y = -5 \end{cases}$.

Solution:

Answer: The system has two solutions: $(0, 5)$ and $(3, 8)$.

Problem 6

Solve: $\begin{cases} y = x^2 + 2 \\ 3x + 4y = -8 \end{cases}$.

Solution:

Answer: There is no solution to the system.

Just for fun, try solving the next problem. It involves a system of equations, but this time both equations are quadratics. I have confidence that you can do the problem without any explanation on my part.

Problem 7

Solve: $\begin{cases} y = x^2 - 6x + 5 \\ y = x^2 - 4x + 4 \end{cases}$.

Solution:

Answer: $\left(\frac{1}{2}, 2\frac{1}{4}\right)$

Geometrically, what did you find in Problem 7?

You should now be able to solve any system of equations containing one linear equation and one quadratic equation. Remember that elimination by substitution is often the most efficient method to solve such systems; briefly stated, the three basic steps involved are:

1. **Substitute into** the linear equation.
2. **Solve.**
3. **Substitute back.**

Before beginning the next unit you should solve the following systems.

EXERCISES

Solve:

1. $\begin{cases} y = 3x^2 - 7x + 11 \\ x = 2 \end{cases}$

2. $\begin{cases} y = 4x^2 \\ -4x + y = 0 \end{cases}$

3. $\begin{cases} y = x^2 + 5x \\ y = -2x + 44 \end{cases}$

4. $\begin{cases} y = -x^2 - 1 \\ -x + 5y = -5 \end{cases}$

5. $\begin{cases} y = x^2 + 5x - 21 \\ y = x \end{cases}$

6. $\begin{cases} y = x^2 + 2x + 5 \\ -4x + 2y = -7 \end{cases}$

7. $\begin{cases} y = 2x^2 - 4x \\ y = 8x - 18 \end{cases}$

8. $\begin{cases} y = 3x^2 + 8x - 2 \\ y = 1 \end{cases}$

9. $\begin{cases} y = 10x^2 - x - 3 \\ y = x + 20 \end{cases}$

10. $\begin{cases} 2x - y + 4 = 0 \\ y = \frac{1}{2}x^2 + x + \frac{5}{2} \end{cases}$

11. $\begin{cases} y = x^2 + 13 \\ y = -10x + 12 \end{cases}$

12. $\begin{cases} y = x^2 + 5x + 1 \\ y = x^2 + 3x + 9 \end{cases}$ Here is another one for fun.

UNIT 28

Solving Inequalities— First-Degree

The purpose of this unit is to provide you with an understanding of inequalities. When you have finished the unit, you will be able to solve *all* first-degree inequalities.

INEQUALITY SIGNS

There are two inequality signs:

$<$ read "less than"

$>$ read "greater than"

Sometimes an inequality sign and an equal sign are combined:

$\leq$ read "less than or equal to"

$\geq$ read "greater than or equal to"

For our purposes in this book we will use a "commonsense" definition of $<$, based on your familiarity with the number line. Remember the number line? It looks like this:

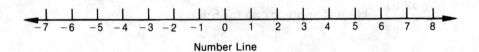

Number Line

> Definition: $a < b$ if a is to the left of b on the number line.

Here are some examples:

$2 < \quad 3$ because 2 is to the left of 3 on the number line.

$-2 < \quad 5$ because -2 is to the left of 5 on the number line.

$-5 < -2$ because -5 is to the left of -2.

$-7 < -6$ because -7 is to the left of -6.

Before proceeding, write a few more examples yourself.

The sign > can be defined in a similar manner. How would the definition read? An alternative approach is to define > as follows:

$$b > a \text{ if and only if } a < b.$$

In words, b is greater than a if a is less than b.

For instance:

$3 > \quad 2$ because $\quad 2 < \quad 3$.

$0 > -1$ because $-1 < \quad 0$.

$-4 > -5$ because $-5 < -4$.

Notice that the inequality sign always opens to the larger number.

Observe that $5 > 2$ has the same meaning as $2 < 5$. Both statements indicate that 5 is the larger number and 2 is the smaller number. Thus:

$3 > -1$ can be rewritten as $-1 < \quad 3$.

$-8 < -2$ can be rewritten as $-2 > -8$.

$3 > \quad x$ can be rewritten as $\quad x < \quad 3$.

$0 < \quad y$ can be rewritten as $\quad y > \quad 0$.

I suggest rewriting most inequalities using <, because it is clearer to visualize the relationship of the numbers and the symbol, reading from left to right, on the number line.

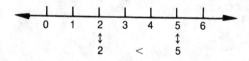

The direction to which an inequality sign points is referred to as its **sense**. For example, $x < 3$ and $a < b$ are said to have the same sense because their symbols are pointing in the same direction. In contrast, $a > c$ and $2 < 5$ are said to be of opposite sense because their symbols are pointing in different directions.

FIRST-DEGREE INEQUALITIES

Remember that in any first-degree equation:

There is only one variable.
It is involved in only the four fundamental operations.
It is never multiplied by itself.
It is never in a denominator.

A **first-degree inequality** has the same characteristics as a first-degree equation except that in place of the equal sign there is an inequality sign.

Here are some examples of first-degree inequalities:

$$2x + 5 < 7$$

$$3(x - 1) + x \geq 2 - x$$

$$1 - x \leq 15(3 + 2x) - x$$

To solve a first-degree inequality is to find the values of x that satisfy the inequality. The basic strategy is the same as that used to solve first-degree equations—get all terms involving x on one side of the inequality sign, and get all other terms on the other side.

To accomplish this, we use two rules:

Rule 1:   **A term may be transposed from one side of the inequality to the other if and only if its sign is changed as it crosses the inequality sign.**

EXAMPLE 1   If $x + 5 < 7$,

then $x < 7 - 5$

Note sign change.

and $x < 2$.

EXAMPLE 2   If $1 - x > -6$,

then $1 + 6 > x$

and $7 > x$

or $x < 7$.

Rule 2:   **Both sides of an inequality may be multiplied or divided by the same nonzero number provided that:**
    a.   **if the number is positive, the direction of the inequality remains the same;**
    b.   **if the number is negative, the direction of the inequality is reversed.**

EXAMPLE 3   If $6 < 15$,

then $\dfrac{6}{3} < \dfrac{15}{3}$

and $2 < 5$.

EXAMPLE 4   If $\frac{1}{4} < 12$,

then $4(\frac{1}{4}) < 4(12)$

and $1 < 48$.

EXAMPLE 5   If $15 > 10$,

then $\dfrac{15}{-5} < \dfrac{10}{-5}$

and $-3 < -2$.

Note: The direction of the **inequality must be reversed** when the inequality is divided by the same negative number.

EXAMPLE 6   If $-2x \le 8$,

then $\dfrac{-2x}{-2} \ge \dfrac{8}{-2}$

and $x \ge -4$.

Note again: The direction of the **inequality must be reversed** when the inequality is divided by the same negative number.

EXAMPLE 7   If $4 > -2$,

then $(-1)4 < (-1)(-2)$

and $-4 < 2$.

Note: The direction of the **inequality must be reversed** when the inequality is multiplied by the same negative number.

Now let's get on with the business of solving first-degree inequalities. The same four steps used to solve first-degree equations will again be used to solve inequalities. However, we must remember to change the direction of the inequality if we multiply or divide by the same negative number.

Recall the four steps:

1. **Simplify:** remove parentheses,
   clear of fractions,
   collect like terms.
2. **Transpose.**
3. **Simplify.**
4. **Divide by coefficient.**

Now, we must remember:

> **Reverse direction** of inequality symbol whenever inequality is multiplied or divided by the same negative number.

Here are some examples that illustrate the use of these steps in solving first-degree inequalities. Read through each step, and be sure you understand what has happened to the terms.

EXAMPLE 8

Solve: $4x - 7 > 6x + 5$.

Solution:

$$4x - 7 > 6x + 5$$
$$4x - 6x > 5 + 7$$
$$-2x > 12$$
$$\frac{-2x}{-2} < \frac{12}{-2} \qquad \text{Reverse direction.}$$
$$x < -6$$

So the answer is $x < -6$ (or $-6 > x$).

To say that $x < -6$ is the answer to the inequality means that all numbers less than -6 satisfy the inequality. For example, if we substitute $x = -7$ into the inequality, we obtain

$$4(-7) - 7 > 6(-7) + 5$$
$$-28 - 7 > -42 + 5$$
$$-35 > -37$$

and, rewriting, $\qquad -37 < -35$

which is true, so we say that $x = -7$ satisfies the inequality.

Graphically, $x < -6$ can be represented on the number line with an open circle at -6 and a heavy line to the left. The open circle indicates that -6 is not part of the answer. The heavy line indicates that all numbers to the left of -6 are part of the answer, including such numbers as -6.1 and -6.25.

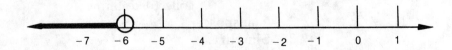

EXAMPLE 9

Solve: $5(x + 3) \geq 31 + x$.

Solution:

$$5(x + 3) \geq 31 + x$$
$$5x + 15 \geq 31 + x$$
$$5x - x \geq 31 - 15$$
$$4x \geq 16$$
$$x \geq 4$$

So the answer is $x \geq 4$ or, rewritten, $4 \leq x$.

To say that $x \geq 4$ is the answer to the inequality means that all numbers greater than or equal to 4 will satisfy the inequality.

If the answer were graphed on a number line, there would be a closed circle at 4 and a heavy line to the right.

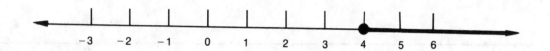

Now try to solve the inequalities in Problems 1 and 2. Remember to **reverse the direction of the inequality symbol** whenever the inequality is multiplied or divided by the same negative number.

Problem 1

Solve: $3x - (x + 7) \geq (x - 2) + 4$.

Solution:

Answer: $x \geq 9$ (or $9 \leq x$)

Problem 2

Solve: $x - 4 - 2(6 - x) > 2(3x - 5)$.

_ _

Solution:

Answer: $x < -2$

_ _

Now, in contrast to our earlier examples, consider the following inequality.

EXAMPLE 10

Solve: $2x + 1 + x < 3(x + 2)$.

_ _

Solution: $2x + 1 + x < 3(x + 2)$

$$1 + 3x < 3x + 6$$
$$3x - 3x < 6 - 1$$
$$0 < 5$$

The solution is the entire set of real numbers.

_ _

The inequality in Example 10 is *always* true, regardless of the value of x, because 0 is always less than 5. Its solution is the entire set of real numbers. In other words, x can equal any number.

For example,   if $x = 0$, then $2(0) + 1 + (0) < 3[(0) + 2]$

and $1 < 6$.

Or: if $x = 5$, then $2(5) + 1 + (5) < 3[(5) + 2]$

and $10 + 1 + 5 < 3(7)$

$$16 < 21.$$

You should now be able to solve any first-degree inequality.

The same four basic steps—simplify, transpose, simplify, and divide—are used to solve first-degree inequalities. In addition, you must remember to reverse the direction of the inequality symbol whenever you multiply or divide an inequality by the same negative number.

Now try to solve the problems in the exercises.

EXERCISES

Solve, and graph the answer for each problem on a number line:

1. $10 + 2x \leq 12$
2. $4 - (12 - 3x) \leq -5$
3. $5x < 22 - (2x + 1)$
4. $4x + (3x - 7) > 2x - (28 - 2x)$
5. $5 - 3x \leq 23$
6. $3x + 4(x - 2) \geq x - 5 + 3(2x - 1)$
7. $3x - 2(5x + 2) > 1 - 5(x - 1) + x$
8. $3x - 2(x - 5) < 3(x - 1) - 2x - 11$
9. $3x + 4(x - 2) + 7 > x - 5 + 3(2x - 1)$
10. $5x - 2(3x - 4) > 4[2x - 3(1 - 3x)]$ Be careful!

_ _

If additional practice is needed:
   Leithold, page 351, problems 1–12
   Lial and Miller, page 65, problems 15–18
   Peters, pages 378 and 379, problems 1–8
   Rees, Sparks, and Rees, page 86, problems 1–20

UNIT 29

Solving Inequalities— Second-Degree

In this unit we will continue our discussion of inequalities. When you have completed the unit, you will be able to solve all second-degree inequalities in one variable.

WRITING DOUBLE INEQUALITIES

A **double inequality** has two inequality symbols of the same sense combined in one statement.

For example, $2 < x < 5$ is a double inequality. In words, $2 < x < 5$ means that x represents all numbers greater than 2 but less than 5. In other words, x lies between 2 and 5 on the number line.

Graphically, $2 < x < 5$ is represented as follows:

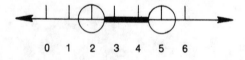

Observe that $5 > x > 2$ has the same meaning as $2 < x < 5$. Both statements indicate that 5 is the larger number, 2 is the smaller number, and x lies between them.

As stated in Unit 28, I prefer to write most inequalities using the $<$ symbol, because it is clearer to visualize the relationship of the numbers and the symbols on the number line. For example:

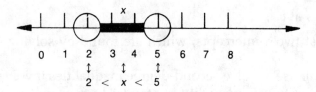

What does $-1 \leq x \leq 4$ mean?

To say that $-1 \leq x \leq 4$ indicates that x represents all numbers between -1 and 4, including the numbers -1 and 4. Graphically it would look like this:

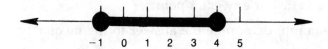

SECOND-DEGREE INEQUALITIES

Recall from Unit 21:

> Definition: $ax^2 + bx + c = 0$, with a, b, and c being real numbers, $a \neq 0$, is called a **second-degree equation** or **quadratic equation.**

A second-degree or quadratic inequality has the same characteristics as a second-degree equation, except that in place of the equal sign there is an inequality symbol. The standard form of the inequality has **all** terms on the left side of the inequality symbol with **only** 0 on the right.

Examples of second-degree inequalities in standard form are:

$$x^2 - 3x + 5 < 0$$

$$7x^2 + 2x - 1 > 0$$

$$x^2 \geq 0$$

$$4x^2 + 7 \leq 0$$

whereas

$$x^3 - x^2 > 0 \text{ is third-degree}$$

$$x^2 + y < 0 \text{ has two variables}$$

$$3x - 7 \leq x \text{ is first-degree}$$

To solve a second-degree or quadratic inequality is to find the values of x that satisfy the inequality. The basic technique developed here will rely heavily on our knowledge of graphing quadratic or second-degree equations in two variables.

Recall some of the basic facts (page 184) about the graph of a quadratic equation in standard form: $y = ax^2 + bx + c$.

1. The graph is a smooth, ⌣-shaped curve called a parabola.

2. If a is positive, the curve opens up. If a is negative, the curve opens down.

3. The y-intercept is at c.

4. There are at most two x-intercepts, which are found by solving $0 = ax^2 + bx + c$.

   The procedure I will use to solve second-degree inequalities involves three major steps. (If necessary, first rewrite the inequality in standard form, that is, put all terms on the left side with only 0 remaining on the right side of the inequality symbol.)

1. Let y = quantity on the left, resulting in a quadratic equation in two variables.

2. Do a rough sketch of the quadratic equation, showing only the intercept(s). There is no need to locate the vertex or any extra points.

3. By inspection of the graph, determine the answer to the inequality.

   The examples that follow illustrate this procedure.

EXAMPLE 1

Solve: $x^2 - 2x - 3 < 0$.

Solution: Let $y = x^2 - 2x - 3$ and graph.

This is a quadratic equation; graph is a parabola.

Since $a = 1$ is positive, curve opens up.

The y-intercept is $c = -3$.

Find the x-intercepts, at most two, by solving:

$$0 = x^2 - 2x - 3$$
$$0 = (x + 1)(x - 3)$$
$$x = -1 \quad \text{or} \quad x = 3$$

The x-intercepts are -1 and 3.

Do a rough sketch of the parabola, showing only the intercepts. There is no need to locate the vertex.

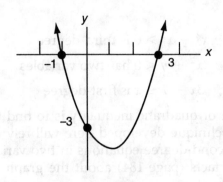

By inspection of the graph, answer this question:

for what values of x is $x^2 - 2x - 3 < 0$?

Or, in other words:

for what values of x is $y < 0$?

Or, in still other words:

for what values of x is the curve below the x-axis?

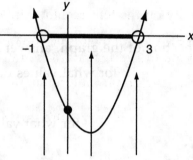

The curve is below the
x-axis when x is any number
between -1 and 3.

Answer: $-1 < x < 3$

To say that $-1 < x < 3$ is the answer to the inequality means that all numbers between -1 and 3 satisfy the inequality. For example, if we substitute $x = 1$ into the inequality, we obtain

$$(1)^2 - 2(1) - 3 < 0$$
$$1 - 2 - 3 < 0$$
$$-4 < 0$$

which is true, so we say that $x = 1$ satisfies the inequality.

EXAMPLE 2

Solve: $x^2 + 5x + 6 < 0$.

Solution: Let $y = x^2 + 5x + 6$ and graph.

This is a quadratic equation; graph is a parabola.

Since $a = 1$ is positive, curve opens up.

The y-intercept is $c = 6$.

Find the x-intercepts, at most two, by solving:

$$0 = x^2 + 5x + 6$$
$$0 = (x + 2)(x + 3)$$
$$x = -2 \quad \text{or} \quad x = -3$$

The x-intercepts are -2 and -3.

Do a rough sketch of the parabola, showing only the intercepts. There is no need to locate the vertex.

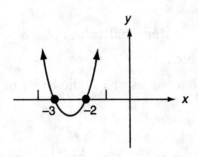

By inspection of the graph, answer this question:

for what values of x is $x^2 + 5x + 6 < 0$?

Or:

for what values of x is $y < 0$?

Or:

for what values of x is the curve below the x-axis?

The curve is below the x-axis when x is any number between -2 and -3.

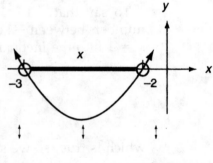

Answer:　$-3 < x < -2$

Try solving this problem.

Problem 1

Solve: $x^2 - 3x - 10 < 0$.

Solution:

Answer: $-2 < x < 5$

I'll do another example; then you try two problems.

EXAMPLE 3

Solve: $-x^2 + 4 < 0$.

Solution: Let $y = -x^2 + 4$ and graph.

Since $a = -1$ is negative, parabola opens down.

The y-intercept is $c = 4$.

Find the x-intercepts, at most two, by solving:

$$0 = -x^2 + 4$$
$$x^2 = 4$$
$$x = -2 \quad \text{or} \quad x = 2$$

The x-intercepts are -2 and 2.

Do a rough sketch of the parabola, showing only intercepts.

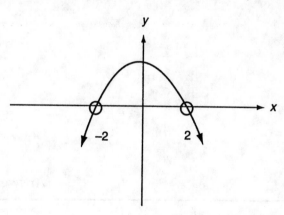

By inspection of the graph, answer this question:

for what values of x is the curve below the x-axis?

In this example, the parabola is opening down, and the two ends of the curve are below the x-axis. Thus the curve is below the x-axis when x is less than -2 or when x is greater than 2.

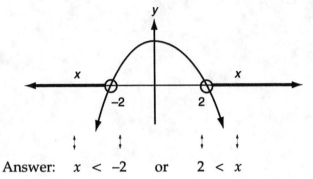

Answer: $x < -2$ or $2 < x$

Note: The answer cannot be written as a double inequality because x does *not* lie between two numbers.

Problem 2

Solve: $-x^2 + 4x + 5 < 0$.

Solution:

Answer: $x < -1$ or $5 < x$

Problem 3

Solve: $x^2 + 1 < 0$.

Solution:

Answer: no solution

The entire graph is above the x-axis or, in other words, $x^2 + 1$ is always positive.

Thus far, all our examples have been inequalities with the $<$ symbol. When an inequality has a $>$ symbol, the procedure for solving it is similar except for the final question.

To solve an inequality, written in standard form, involving $>$, the final question becomes:

for what values of x is $y > 0$?

Or, in other words:

for what values of x is the curve **above** the x-axis?

EXAMPLE 4

Solve: $-x^2 + x + 12 > 0$.

Solution: Let $y = -x^2 + x + 12$ and graph.

This is a quadratic equation; graph is a parabola.

Since $a = -1$ is negative, curve opens down.

The y-intercept is $c = 12$.

The x-intercepts:

$$0 = -x^2 + x + 12$$
$$0 = x^2 - x - 12$$
$$0 = (x - 4)(x + 3)$$
$$x = 4 \quad \text{or} \quad x = -3$$

The x-intercepts are -3 and 4.

A rough sketch of the parabola is:

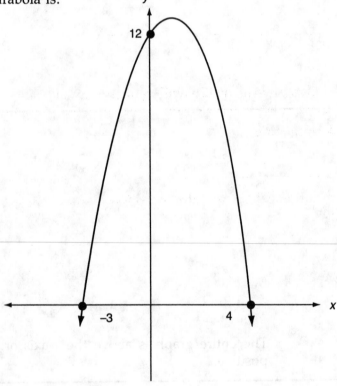

By inspection of the graph, answer this question:

for what values of x is $-x^2 + x + 12 > 0$?

Or:

for what values of x is $y > 0$?

Or:

for what values of x is the curve **above** the x-axis?

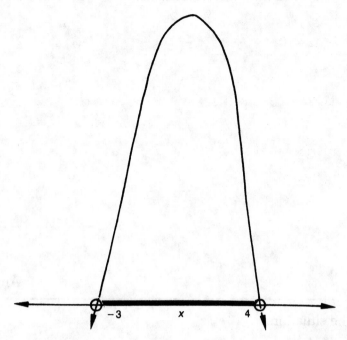

Answer: $-3 < x < 4$

Here is an inequality with > for you to solve.

Problem 4

Solve: $-x^2 + 7x - 6 > 0$.

Solution (There is additional space on page 246.):

Answer: $1 < x < 6$

To generalize the situation:

> Given $ax^2 + bx + c < 0$ to solve, the final question to be answered is:
>
> for what values of x is the curve **below** the x-axis?
>
> Given $ax^2 + bx + c > 0$ to solve, the final question to be answered is:
>
> for what values of x is the curve **above** the x-axis?

Here is one more example in detail with explanation.

EXAMPLE 5

Solve: $3x^2 \geq -4x - 1$.

Solution: Write in standard form: $3x^2 + 4x + 1 \geq 0$.

Let $y = 3x^2 + 4x + 1$ and graph.

This is a quadratic equation; graph is a parabola.

Since $a = 3$ is positive, curve opens up.

The y-intercept is $c = 1$.

The x-intercepts:

$$0 = 3x^2 + 4x + 1$$
$$0 = (3x + 1)(x + 1)$$
$$x = -\tfrac{1}{3} \quad \text{or} \quad x = -1$$

The x-intercepts are -1 and $-\tfrac{1}{3}$.

A rough sketch of the parabola is:

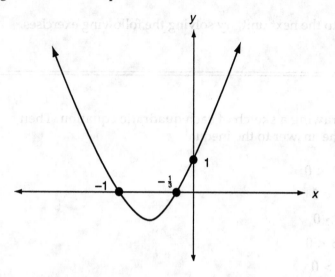

By inspection of the graph, answer this question:

for what values of x is the curve **above** or **on** the x-axis?

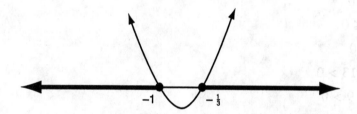

Answer: $x \leq -1$ or $-\frac{1}{3} \leq x$

Note: The x-intercepts are part of the answer. Why?

Convince yourself, by substitution, that both $x = -1$ and $x = -\frac{1}{3}$ satisfy the inequality.

_____

You should now be able to identify and solve second-degree or quadratic inequalities in one variable. Remember our three-step procedure, which, simply stated, is:

1. Let y = quantity on left.
2. Do a rough sketch showing only intercept(s).
3. By inspection, determine the answer.

Also remember:

If the inequality involves $<$, the final question to be answered is:

for what values of x is the curve **below** the x-axis?

If the inequality involves $>$, the final question to be answered is:

for what values of x is the curve **above** the x-axis?

If the inequality is combined with an = sign, the x-intercepts are part of the answer.

Before going on to the next unit, try solving the following exercises.

EXERCISES

Solve for x by drawing a sketch of each quadratic equation. Then by inspection of the graph, determine the answer to the inequality.

1. $x^2 + 4x - 21 < 0$
2. $x^2 + 2x - 8 < 0$
3. $x^2 - x - 12 > 0$
4. $-x^2 - 4x - 3 < 0$
5. $-x^2 - x + 2 > 0$
6. $-4x^2 \geq 0$
7. $x^2 - 4x + 4 \leq 0$
8. $4 + 2x^2 \leq 0$
9. $9x^2 - 9x \leq 0$
10. $25 - x^2 \geq 0$
11. $x^2 - 4x - 13 > 0$
12. $x^2 + 6x + 9 > 0$
13. $-x^2 + 2x + 15 \leq 0$
14. $2x^2 + 4x - 1 < 0$
15. $2x^2 + 1 \leq 0$
16. $-x^2 + 10x \leq 0$
17. $2x^2 + 3x - 5 \leq 0$
18. $3x^2 - 3x + 2 > 0$
19. $-6x^2 - x - 2 \geq 0$
20. $2x^2 - 12x + 3 \leq 0$

If additional practice is needed:
   Leithold, page 358, problems 1–11
   Lial and Miller, page 89, problems 1–18
   Peters, page 379, problems 9–14

UNIT 30

Logarithms

The purpose of this unit is to provide you with a brief overview of logarithms. First the definition and notation used with logarithms will be introduced. Then you will learn about logarithmic properties and the ways these properties are used to simplify expressions involving logarithms.

The equation $\log_b N = x$ is read "the logarithm of N to the base b is x." Often "the logarithm of N" is shortened and read as "$\log N$."

Definition: x is called the **logarithm of N to the base b** if $b^x = N$, where N and b are both positive numbers, $b \neq 1$.

In other words:

$$\log_b N = x \quad \text{if and only if} \quad b^x = N.$$

Examples of logarithms are:

$$\log_3 9 = 2 \quad \text{because} \quad 3^2 = 9$$

$$\log_2 8 = 3 \quad \text{because} \quad 2^3 = 8$$

$$\log_7 7 = 1 \quad \text{because} \quad 7^1 = 7$$

$$\log_5 25 = 2 \quad \text{because} \quad 5^2 = 25$$

$$\log_{13} 1 = 0 \quad \text{because} \quad 13^0 = 1$$

$$\log_2 \tfrac{1}{2} = -1 \quad \text{because} \quad 2^{-1} = \tfrac{1}{2}$$

LOGARITHMIC AND EXPONENTIAL FORMS

Notice that **the logarithm of a positive number N is the exponent to which the base must be raised to produce the number N.**

$$\log_b N = x \text{ is called the logarithmic form,}$$
$$b^x = N \text{ is called the exponential form,}$$
and the two statements are equivalent.

Try to develop your ability to go from one form to the other.

EXAMPLE 1

Express $\log_8 16 = \frac{4}{3}$ in exponential form.

Solution: If $\log_8 16 = \frac{4}{3}$
then $8^{4/3} = 16.$

Cover the answers, and then try to express each of the following in exponential form.

Problem 1 $\log_3 243 = 5$

Problem 2 $\log_2 16 = 4$

Problem 3 $\log_{11} 121 = 2$

Problem 4 $\log_u w = y$

Answers: 1. $3^5 = 243$ 2. $2^4 = 16$ 3. $11^2 = 121$ 4. $u^y = w$

Students usually find it more difficult to go from exponential form to logarithmic form. I start by writing the base first, then the exponent. Remember that the logarithm of a number is an **exponent.**

EXAMPLE 2

Express $5^3 = 125$ in logarithmic form.

Solution: If $5^3 = 125,$
then $\log_5 125 = 3.$

Cover the answers, and then try to express each of the following in logarithmic form.

Problem 5 $4^2 = 16$

Problem 6 $12^0 = 1$

Problem 7 $4^{-5/2} = \frac{1}{32}$

Problem 8 $A^2 = c$

Answers: 5. $\log_4 16 = 2$ 6. $\log_{12} 1 = 0$ 7. $\log_4 \frac{1}{32} = -\frac{5}{2}$ 8. $\log_A c = 2$

- -

Actually any positive number can be used for the base. Logarithms to the base 10 are called **common logs**, and the 10 is omitted from the logarithmic notation. In other words, if no base is written using logarithmic notation, the base is understood to be 10.

EXAMPLE 3

Express $\log A = c$ in exponential form.

- -

Solution: $\log A = c$ denotes a common logarithm with base 10 because no base number is indicated.

Therefore,   if $\log A = c$,

then $10^c = A$.

- -

It's your turn now.

Problem 9

Express $\log 100 = 2$ in exponential form.

- -

Solution:

Answer: $10^2 = 100$

- -

SOLVING LOGARITHMIC EQUATIONS

Now let's try solving some logarithmic equations of the form $x = \log_b N$, where we are to find the missing number.

If you are more comfortable with the exponential form at this stage, I suggest changing from the logarithmic form to the exponential form to solve the equations. Eventually, though, you will need to solve the equations in their original form.

EXAMPLE 4

Solve: $\log_7 7 = x$.

Solution: If $\log_7 7 = x$,

then $7^x = 7$

and $7^x = 7^1$ since $7 = 7^1$.

Answer: $x = 1$

Here's an alternative method, which I think is easier:

In words, $\log_7 7$, means "7 raised to the power that equals 7."

The answer is 1.

EXAMPLE 5

Solve: $\log_2 32 = x$.

Solution: If $\log_2 32 = x$,

then $2^x = 32$

and $2^x = 2^5$ since $32 = 2^5$.

Answer: $x = 5$

Alternative method:

In words, $\log_2 32$ means "2 raised to the power that equals 32."

The answer is 5.

Try Problems 10–12 on your own.

Problem 10

Solve: $\log_8 64 = x$.

Solution:

Alternative method:

In words, $\log_8 64$ means "8 raised to the power that equals 64."

Answer: $x = 2$

Problem 11

Solve: $\log 1000 = x$.

Solution:

Answer: $x = 3$

Problem 12

Solve: $\log_7 1 = x$.

Solution:

Answer: $x = 0$

In the preceding examples and problems, the base and number were given and we had to find the logarithm. The following examples illustrate the technique to be used when the log is given and we wish to solve for either the base or the number.

EXAMPLE 6

Solve: $\log_3 x = 4$.

Solution: If $\log_3 x = 4,$

then $3^4 = x$

and $81 = x$ since $3^4 = 81$.

Answer: $x = 81$

EXAMPLE 7

Solve: $\log_x 9 = \frac{1}{2}$.

Solution: If $\log_x 9 = \frac{1}{2},$

then $x^{1/2} = 9$

or $\sqrt{x} = 9$.

Answer: $x = 81$

Since Example 7 may be a bit confusing, let's do one more before you try Problems 13–17.

EXAMPLE 8

Solve: $\log_x 64 = 6$.

Solution: If $\log_x 64 = 6,$

then $x^6 = 64$

and $x^6 = 2^6$ since $64 = 2^6$.

Answer: $x = 2$

Problem 13

Solve: $\log_5 x = 1$.

Solution:

Answer: $x = 5$

Problem 14

Solve: $\log x = 4$.

Solution:

Answer: $x = 10{,}000$

Problem 15

Solve: $\log_3 x = -2$.

Solution:

Answer: $x = \frac{1}{9}$

Problem 16

Solve: $\log_x 100 = 2$.

Solution:

Answer: $x = 10$

Problem 17

Solve: $\log_x 8 = 3$.

Solution:

Answer: $x = 2$

By now you should be asking such questions as:

These problems were okay, but what if the answer is not obvious?
What if I have something other than an integer value for the log?
What if the number is not an integral power of the base?

The next example is intended to answer most of your questions.

EXAMPLE 9

Solve: $\log 5 = x$.

Solution:　　If　　$\log 5 = x$,

then　　$10^x = 5$.

Problems like this, where x is not an integral value, cannot be solved by elementary algebra. It is possible to find a decimal approximation for the common log of any positive number through the use of prepared tables. However, problems of this type are beyond the intent of this unit and therefore will not be considered.

SIMPLIFYING EXPRESSIONS WITH LOGARITHMS

To accomplish simplification, we have **four** basic properties of logarithms.

These properties are used to shorten computations or to simplify complicated expressions involving products, quotients, powers, and roots.

Properties of Logarithms

Property 1: Product $\log_b AC = \log_b A + \log_b C$

Property 2: Quotient $\log_b \dfrac{A}{C} = \log_b A - \log_b C$

Property 3: Power $\log_b A^k = k \log_b A$

Property 4: Root $\log_b \sqrt[k]{A} = \dfrac{1}{k} \log A$

The following examples and problems illustrate how these four properties can be used to shorten computations. Remember that the logarithm of a positive number N is the exponent to which the base must be raised to produce the number N.

Property 1: Product $\log_b AC = \log_b A + \log_b C$

Verbally, property 1 states that the logarithm of a product of two numbers is equal to the sum of the logarithms of the numbers.

EXAMPLE 10

Find: $\log_3 (81 \cdot 9)$.

Solution: $\log_3 (81 \cdot 9) = \log_3 81 + \log_3 9$

$= 4 + 2$

$= 6$

Answer: $\log_3 (81 \cdot 9) = 6$

In case you are not convinced:

If $\log_3 (81 \cdot 9) \stackrel{?}{=} 6,$

then $3^6 \stackrel{?}{=} (81 \cdot 9)$

and $729 = 729.$

EXAMPLE 11

Question: If log 3 = 0.477 and log 2 = 0.301, what is log 6?

Solution: log 6 = log (3 · 2)

$$= \log 3 + \log 2$$
$$= 0.477 + 0.301$$
$$= 0.778$$

Answer: log 6 = 0.778

Problem 18

Find: $\log_2 (8 \cdot 4)$.

Solution: Use property 1.

Answer: 5

Problem 19

Find: $\log_2 (64 \cdot 32)$.

Solution: Use property 1.

Answer: 11

> Property 2: Quotient $\quad \log_b \dfrac{A}{C} = \log_b A - \log_b C$

Verbally, property 2 states that the logarithm of a quotient of two numbers is equal to the difference of the logarithms of the numbers.

EXAMPLE 12

Find: $\log_3 \frac{1}{27}$.

Solution: $\log_3 \frac{1}{27} = \log_3 1 - \log_3 27$

$$= 0 - 3$$

$$= -3$$

Answer: $\log_3 \frac{1}{27} = -3$

If you are not convinced, change from the logarithmic form to the exponential form and verify that both sides represent the same number.

EXAMPLE 13

Question: If $\log 3 = 0.477$ and $\log 2 = 0.301$, what does $\log 1.5$ equal?

Solution: $\log 1.5 = \log \frac{3}{2}$

$$= \log 3 - \log 2$$

$$= 0.477 - 0.301$$

$$= 0.176$$

Answer: $\log 1.5 = 0.176$

Problem 20

Find: $\log_2 \frac{4}{64}$.

Solution: Use property 2.

Answer: -4

Problem 21

Find: $\log_2 \frac{1}{4}$.

Solution: Use property 2.

Answer: -2

Property 3: Power $\qquad \log_b A^k = k \log_b A$

Verbally, property 3 states that the logarithm of a power of a number is equal to the power times the logarithm of the number.

EXAMPLE 14

Find: $\log_3 (81)^5$.

Solution: $\log_3 (81)^5 = 5 \log_3 81$

$\qquad\qquad\qquad = 5(4)$

$\qquad\qquad\qquad = 20$

Answer: 20

$$\boxed{\text{Property 4: } \log_b \sqrt[k]{A} = \frac{1}{k}\log_b A}$$

Property 4 follows directly from property 3. Verbally, property 4 states that to take the logarithm of a root of a number, rewrite the root of the number using a fractional exponent and then apply property 3.

The following example is an illustration of the use of this property.

EXAMPLE 15

Find: $\log_3 \sqrt{27}$.

- -

Solution: $\log_3 \sqrt{27} = \log_3 (27)^{1/2}$

$$= \frac{1}{2}\log_3 27$$

$$= \frac{1}{2}(3)$$

$$= 1.5$$

Answer: 1.5

- -

Problem 22

Find: $\log_2 (32)^3$.

- -

Solution: Use property 3.

Answer: 15

- -

Problem 23

Find: $\log_2 \sqrt{32}$.

- -

Solution: Use property 4.

Answer: 2.5

- -

As previously stated, the four basic properties of logarithms are used also to simplify complicated expressions involving products, quotients, powers, and roots.

Let's try some dandy examples and problems.

EXAMPLE 16

Simplify: $\log \dfrac{xy}{z}$.

Solution: $\log \dfrac{xy}{z} = \log xy - \log z$ prop. 2

$= \log x + \log y - \log z$ prop. 1

EXAMPLE 17

Simplify: $\log \dfrac{x^3 y}{z}$.

Solution: $\log \dfrac{x^3 y}{z} = \log x^3 y - \log z$ prop. 2

$= \log x^3 + \log y - \log z$ prop. 1

$= 3 \log x + \log y - \log z$ prop. 3

EXAMPLE 18

Simplify: $\log \sqrt{xy}$ with $x > 0$ and $y > 0$.

Solution: $\log \sqrt{xy} = \log (xy)^{1/2}$

$= \tfrac{1}{2} \log xy$ prop. 4

$= \tfrac{1}{2}(\log x + \log y)$ prop. 1

$= \tfrac{1}{2} \log x + \tfrac{1}{2} \log y$

Problem 24

Simplify: $\log x^2 y$.

Solution:

Answer: $2 \log x + \log y$

Problem 25

Simplify: $\log \dfrac{xy}{z}$.

Solution:

Answer: $\log x + \log y - \log z$

Problem 26

Simplify: $\log \sqrt{\dfrac{x}{y}}$ with $x > 0$ and $y > 0$.

Solution:

Answer: $\tfrac{1}{2}(\log x - \log y)$

You should now have a basic understanding of logarithms. Remember that the logarithm of a positive number N is the exponent to which the base must be raised to produce the number N.

Also, you should be familiar with both logarithmic notation and exponential notation and be able to go from one form to the other.

Finally, you should be able to simplify logarithmic expressions involving products, quotients, and powers using the basic properties of logarithms.

Before beginning the next unit you should work the following exercises.

EXERCISES

Solve for x:

1. $\log_3 81 = x$
2. $\log_5 125 = x$
3. $\log_7 \left(\frac{1}{7}\right) = x$
4. $\log 1 = x$
5. $\log_3 x = 2$

6. $\log_7 x = 0$
7. $\log_9 x = \frac{1}{2}$
8. $\log_x 27 = 3$
9. $\log_x 49 = 2$
10. $\log_x 121 = 2$

If $\log 3 = 0.477$ and $\log 2 = 0.301$, find the following:

11. $\log 9$
12. $\log 8$
13. $\log 12$ (Hint: $12 = 2^2 \cdot 3$.)
14. $\log \frac{2}{3}$

15. $\log \frac{1}{2}$
16. $\log \sqrt{3}$
17. $\log \sqrt[3]{2}$

Express each of the given logarithms in terms of logarithms of x, y, and z, where the variables are all positive.

18. $\log x^5$

19. $\log 2xy^3$

20. $\log \dfrac{x^2}{y}$

21. $\log \dfrac{x}{yz}$ Be careful!

22. $\log \sqrt{x^3 y}$

If additional practice is needed:
Leithold, page 274, problems 1–42, and page 281, problems 1–38
Lial and Miller, pages 112 and 113, problems 1–18
Peters, page 433, problems 1–18, and pages 434 and 435, problems 1–24
Rees, Sparks, and Rees, pages 398 and 399, problems 1–64

UNIT 31

Right Triangles

The purpose of this, the last unit, is to provide you with a working knowledge of two frequently encountered triangles—the 30°-60°-90° triangle and the isosceles right triangle. When you have finished this unit, you will be able to recognize each of these triangles and, given the length of the one side, to find the lengths of the other two sides.

You probably recall that triangles are labeled using three capital letters, one at each vertex. The triangle shown below is referred to as triangle ABC or, simply, $\triangle ABC$. The sides are labeled using lower case letters as follows:

The side opposite angle A is a.
The side opposite angle B is b.
The side opposite angle C is c.

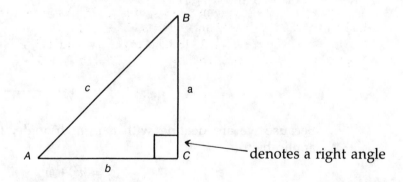

In any triangle the sum of the angles is equal to 180°. An angle of 90° is called a right angle. A **right triangle** is a triangle with a 90° angle. Thus triangle ABC is a right triangle.

265

In a right triangle the side opposite the right angle is called the **hypotenuse.** In triangle ABC, angle C is the right angle and c is the hypotenuse. The hypotenuse is longer than either of the other two sides.

One of the most useful theorems with regard to right triangles is the Pythagorean theorem. More than likely, you can recall some version of it yourself.

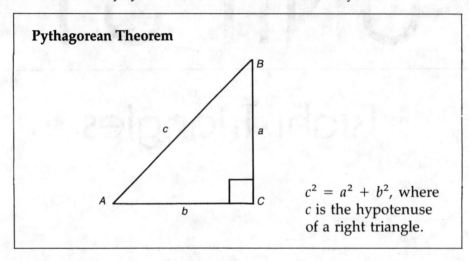

Pythagorean Theorem

$c^2 = a^2 + b^2$, where c is the hypotenuse of a right triangle.

Or, in words, given a right triangle, the square of the length of the hypotenuse is equal to the sum of the squared lengths of the sides.

Be careful. The Pythagorean theorem is applicable only to right triangles.

EXAMPLE 1

Question: Given a right triangle with sides of 5 and 12, what is the length of the hypotenuse?

Solution: Begin by drawing a picture.

Label what is given in the problem.

Let $a = 5$ and $b = 12$.

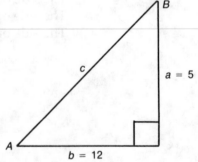

Because we are dealing with a right triangle, the Pythagorean theorem is applicable:

$$c^2 = a^2 + b^2$$
$$= (5)^2 + (12)^2$$
$$= 25 + 144$$
$$= 169$$

$$c = \sqrt{169}$$
$$= 13$$

Answer: The length of the hypotenuse is 13.

Note: Only the positive square root of 169 is the answer, because c represents the length of a side of a triangle. A negative number would be meaningless for an answer in this situation.

EXAMPLE 2

Question: The hypotenuse of a right triangle is 3, and a side adjacent to it is 2. What is the length of the third side?

Solution: Begin by drawing a picture.

Label what is given in the problem.

Let $c = 3$, and $a = 2$.

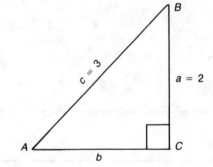

By the Pythagorean theorem:

$$c^2 = a^2 + b^2$$
$$(3)^2 = (2)^2 + b^2$$
$$9 = 4 + b^2$$
$$5 = b^2$$
$$\sqrt{5} = b$$

Answer: The length of the third side of the triangle is $\sqrt{5}$ or, if an approximation is sufficient, 2.236.

I am sure you are ready to try a few problems now.

Problem 1

Question: One side of a right triangle is 7, and the other side is 5. What is the length of the hypotenuse?

_____

Solution:

Answer: $\sqrt{74}$

_____

Problem 2

Question: The hypotenuse of a right triangle is 5, and an adjacent side is 3. What is the length of the third side?

_____

Solution:

Answer: 4

_____

Notice that, when you are given **any two sides of a right triangle,** the third side can be found by using the Pythagorean theorem.

Problem 3

Question: The longest side of a right triangle is 11, and the shortest side is 3. What is the length of the other side?

- -

Solution:

Answer: $\sqrt{112}$

- -

30°-60°-90° TRIANGLES

One of the most frequently encountered right triangles is the **30°-60°-90° triangle,** so named because

one angle is 30°,
another angle is 60°, and
the third angle is 90°.

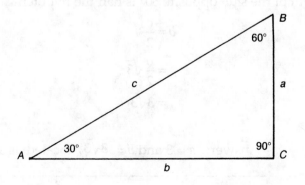

In a 30°-60°-90° triangle the sides have the following relationships:

> The length of the **side opposite the 30° angle** is equal to half the length of the hypotenuse, or, stated as a formula,
>
> $$a = \frac{c}{2}$$
>
> The length of the **side opposite the 60° angle** is equal to half the length of the hypotenuse times $\sqrt{3}$, or, stated as a formula,
>
> $$b = \frac{c}{2}\sqrt{3}$$

EXAMPLE 3

Question: If the hypotenuse of a 30°-60°-90° triangle is 6, what are the lengths of the other sides?

Solution: Begin by drawing a picture.

Label what is given in the problem.

Given $c = 6$.

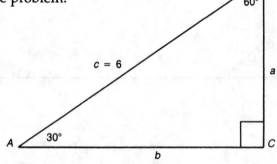

The length of the side opposite 30° is half the hypotenuse.

$$a = \frac{c}{2}$$

$$= \frac{6}{2}$$

$$= 3$$

The length of the side opposite 60° is half the hypotenuse times $\sqrt{3}$.

$$b = \frac{c}{2}\sqrt{3}$$

$$= \frac{6}{2}\sqrt{3}$$

$$= 3\sqrt{3}$$

Answer: $a = 3$ and $b = 3\sqrt{3}$, located as shown in the drawing

Try Problems 4–6.

Problem 4

Question: Given the drawing at the right,
what are the lengths of a and b?

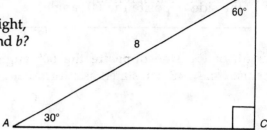

Solution:

Answer: $a = 4$ and $b = 4\sqrt{3}$

Now suppose we change the conditions around a bit.

Problem 5

Question: If the length of the side opposite a 30° angle in a right triangle is 10, what is the
length of the hypotenuse?

Solution:

Answer: $c = 20$

Problem 6

Question: If the side opposite a 30° angle in a right triangle is 17, what is the length of the side opposite the 60° angle?

Solution:

Answer: $17\sqrt{3}$

In the next example I will do the more difficult situation where the length of the side opposite the 60° angle is given.

EXAMPLE 4

Question: If the side opposite the 60° angle in a right triangle is 8, what are the lengths of the other two sides?

Solution: We are given $b = 8$.

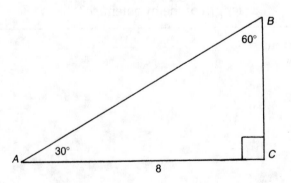

And we know that the side opposite the 60° angle is equal to half the hypotenuse times $\sqrt{3}$, or

$$b = \frac{c}{2}\sqrt{3}$$

So, by substitution,

$$8 = \frac{c}{2}\sqrt{3}$$

$$16 = c\sqrt{3}$$

$$\frac{16}{\sqrt{3}} = c$$

Or, if you prefer the denominator rationalized,

$$c = \frac{16\sqrt{3}}{\sqrt{3}\cdot\sqrt{3}}$$

$$= \frac{16\sqrt{3}}{3}$$

Once the hypotenuse is found, the side opposite the 30° angle is simply half of it.

$$a = \frac{8}{\sqrt{3}} \quad \text{or, if you prefer,} \quad \frac{8\sqrt{3}}{3}$$

Answer: The hypotenuse is $16/\sqrt{3}$, and the other side is $8/\sqrt{3}$.

Here's a similar one for you.

Problem 7

Question: If the side opposite the 60° angle of a right triangle is 2, what are the lengths of the other two sides?

Solution:

Answer: The hypotenuse is $4/\sqrt{3}$ or $4\sqrt{3}/3$, and the short side is $2/\sqrt{3}$ or $2\sqrt{3}/3$.

ISOSCELES RIGHT TRIANGLES

A triangle with two equal sides is called an **isosceles triangle.**
   Another frequently encountered right triangle is the **isosceles right triangle,** so named because

two sides are equal, and
the angle between them is 90°.

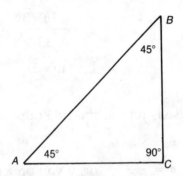

In an isosceles right triangle, if c is the hypotenuse, $a = b$ and angle A = angle B = 45°. The two equal angles of an isosceles triangle are often referred to as the **base angles.**
   As in the 30°-60°-90° triangle, the sides of an **isosceles right triangle** are related.
   In an isosceles right triangle the sides have the following relationships:

The **sides opposite the 45° angles** are equal
and
each side is equal to half the hypotenuse times $\sqrt{2}$, or, stated as a formula,

$$b = \frac{c}{2}\sqrt{2}, \quad \text{with } a = b$$

EXAMPLE 5

Question: If an isosceles right triangle has a hypotenuse of 20, what are the lengths of the other sides?

Solution: As before, begin by drawing a picture.

Label what is given in the problem.

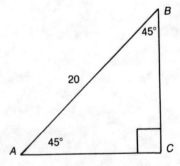

Because this is an isosceles right triangle, angle A and angle B are both 45° and $a = b$. Also, $c = 20$.

In an isosceles right triangle, the side opposite the 45° angle is half the hypotenuse times $\sqrt{2}$; therefore

$$b = \frac{c}{2}\sqrt{2}$$

and, by substitution,

$$b = \frac{20}{2}\sqrt{2}$$

$$= 10\sqrt{2}$$

Answer: The two equal sides are $10\sqrt{2}$.

This time you try the harder problem.

Problem 8

Question: If the side opposite a 45° angle in a right triangle is 5, what is the length of the hypotenuse?

Solution:

Answer: $c = \dfrac{10}{\sqrt{2}}$ or $5\sqrt{2}$

Did you observe that, when working with either the 30°-60°-90° triangle or the isosceles right triangle, if you are given only the length of one side it is possible to find the lengths of the other two? In contrast, the Pythagorean theorem requires that two sides be known in order to find the third.

The preceding examples have been fairly easy, as you probably guessed, and the problems that you are likely to encounter rarely appear in such a straightforward manner. Instead, the information is contained in a hypothetical word problem depicting some real-life situation. However, if you remember to draw a picture and label what is given in the problem, solving it should not be any more difficult than solving the problems you have done already.

A few word problems follow to illustrate what I mean.

EXAMPLE 6

A ladder leans against the side of the building with its foot 7 ft from the building. If the ladder makes an angle of 60° with the ground, how long is the ladder?

Solution: Begin by drawing a picture.

Label what is given in the problem.

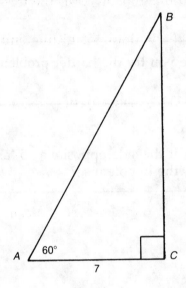

Angle $A = 60°$ and $b = 7$.

Therefore angle B must be 30°. To find the length of the ladder, we apply our knowledge that the side opposite the 30° angle is half the hypotenuse. Since $b = 7$, c must be 14.

Answer: The ladder is 14 ft long.

EXAMPLE 7

When the sun is 45° above the horizon, how long is the shadow cast by a tree 40 ft high?

Solution: We are given angle $A = 45°$ and $a = 40$.

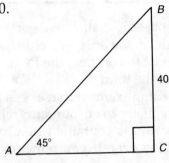

Since this is a right triangle, angle $B = 45°$ and we have an isosceles right triangle. Therefore b must be 40.

Answer: The shadow is 40 ft long.

EXAMPLE 8

From the top of a tree, 25 ft above the ground, the angle of depression to an observer on the ground in 30°. How far is the tree from the observer?

Solution: I will draw the picture and label what is given.

30° = angle of depression

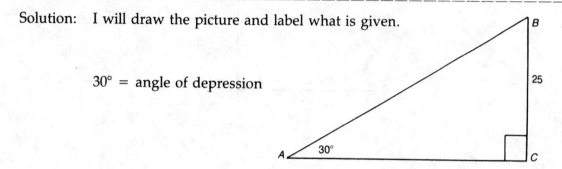

You can finish the problem from here.

Answer: The tree is $25\sqrt{3}$ ft away.

EXAMPLE 9

Find the height of a street lamp if the angle of elevation of its top is 45° to an observer on the ground at a distance of 25 ft from its base.

Solution: Again I will draw the picture and label what is given; then you finish the problem.

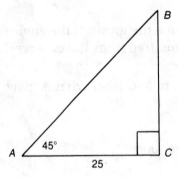

Answer: The street lamp is 25 ft high.

Try Problem 9 on your own without any clues.

Problem 9

A man drives 1000 ft along a road that is inclined 30° to the horizontal. How high above his starting point is he?

Solution:

Answer: He is 500 ft above his starting point.

You probably did fine on that problem, but let's do one more.

Problem 10

A flagpole broken over by the wind forms a right triangle with the ground. If the broken part makes an angle of 45° with the ground, and if the top of the flagpole is now 10 ft from its base, how tall was the flagpole?

Solution:

Answer: The flagpole was $10 + 10\sqrt{2}$ ft tall.

You should now be able to define and label a right triangle. Remember that, when you are given any two sides of a right triangle, the third side can be found by using the Pythagorean theorem, $c^2 = a^2 + b^2$, where c is the hypotenuse.

In addition you should be able to recognize two well-known right triangles.

1.   **The 30°-60°-90° triangle** with properties:
   a.   the side opposite the 30° angle is equal to half the hypotenuse;
   b.   the side opposite the 60° angle is equal to half the hypotenuse times $\sqrt{3}$.

2.   **The isosceles right triangle** with properties:
   a.   the sides opposite the 45° angles are equal;
   b.   the sides opposite the 45° angles are equal to half the hypotenuse times $\sqrt{2}$.

Remember that, when working with either a 30°-60°-90° triangle or an isosceles right triangle, you need to be given only the length of one side in order to find the lengths of the other two sides. Also remember that the suggested procedure when solving such problems is to begin by drawing a picture; then label what is given in the problem.

Now try the following exercises. Full solutions are given at the end of the book for all of these exercises.

EXERCISES

Solve:

1.   The hypotenuse of a right triangle is 7, and a side adjacent to it is 3. Find the length of the third side.

2.   Two sides of a triangle are 1 and 2. Find the length of the hypotenuse.

3.   If the hypotenuse of a 30°-60°-90° triangle is 10, find the lengths of the other sides.

4.   If the side opposite the 60° angle in a right triangle is 5, what are the lengths of the other two sides?

5.   If the side opposite the 30° angle in a right triangle is 1, what are the lengths of the other two sides?

6.   If the side opposite the 45° angle in a right triangle is 6, what are the lengths of the other two sides?

7.   If the hypotenuse of an isosceles right triangle is 10, what are the lengths of the other two sides?

8.   A ladder 20 ft long is resting against a house. If it makes an angle of 45° with the ground, how high up does it reach?

9.   An observer notes that the angle of elevation of the top of a neighboring building is 60°. If the distance from the observer to the neighboring building is 40 ft, how tall is the building?

10.   When the sun is 30° above the horizon, an observer's shadow is 9 ft. How tall is the observer?

Answers to Exercises

Note: For all units except Unit 1, solutions are given in detail for even-numbered problems. For Units 14, and 17, 30, and 31 detailed solutions are provided for all exercises.

UNIT 1

1. -7
2. 4
3. -9
4. -1
5. -48

6. -18
7. 44
8. -4
9. 5
10. 1

11. -1
12. -3
13. 9
14. -3
15. -6

16. -10
17. -24
18. 32
19. 0
20. 1

UNIT 2

1. $9x + 3y + 1$

2. $(3x + 5xy + 2y) + (4 - 3xy - 2x)$
   $3x + 5xy + 2y + 4 - 3xy - 2x$
   $x + 2xy + 2y + 4$

3. $4y + a + 8$

4. $3(2a - b) - 4(b - 2a)$
   $6a - 3b - 4b + 8a$
   $14a - 7b$

5. $14x + y - 17$

6. $5 - 2(x + 2[3 + x])$

   $5 - 2(x + 6 + 2x)$

   $5 - 2x - 12 - 4x$

   $-6x - 7$

7. $-3x + 2$

8. $4x - \{3x - 2[y - 3(x - y)] + 4\}$

   $4x - \{3x - 2[y - 3x < 3y)] + 4\}$

   $4x - \{3x - 2y + 6x - 6y + 4\}$

   $4x - 3x + 2y - 6x + 6y - 4$

   $-5x + 8y - 4$

9. -1

10. $-3a - 5\{2a - 2[2a - (4a - 2) - 2(7 - 3a)]\}$

    $-3a - 5\{2a - 2[2a - 4a + 2 - 14 + 6a]\}$

    $-3a - 5\{2a - 4a + 8a - 4 + 28 - 12a\}$

    I would combine like terms before continuing.

    $-3a - 5\{-6a + 24\}$

    $-3a + 30a - 120$
    $27a - 120$

UNIT 3

1. $x = 2$

2. $2(x+1) - 3(4x-2) = 6x$

$$2x + 2 - 12x + 6 = 6x$$

$$-10x + 8 = 6x$$

$$8 = 6x + 10x$$

$$8 = 16x$$

$$\frac{8}{16} = \frac{16x}{16}$$

$$\frac{1}{2} = x$$

3. $x = 23$

4. $$20 - \frac{3x}{5} = x - 12$$

$$5\left(20 - \frac{3x}{5} = x - 12\right)$$

$$100 - \frac{5 \cdot 3x}{5} = 5x - 60$$

$$100 - 3x = 5x - 60$$

$$-3x - 5x = -60 - 100$$

$$-8x = -160$$

$$\frac{-8x}{-8} = \frac{-160}{-8}$$

$$x = 20$$

5. $x = 2$

6. $$2(x+2) = 5 + \frac{x+1}{3}$$

$$2x + 4 = 5 + \frac{x+1}{3}$$

$$3\left(2x + 4 = 5 + \frac{x+1}{3}\right)$$

$$6x + 12 = 15 + \frac{3(x+1)}{3}$$

$$6x + 12 = 15 + x + 1$$

$$6x - x = 16 - 12$$

$$5x = 4$$

$$\frac{5x}{5} = \frac{4}{5}$$

$$x = \frac{4}{5}$$

7. $x = 6$

8. $$3 - \frac{5(x-1)}{2} = x$$

$$3 - \frac{5x - 5}{2} = x$$

Notice that the minus sign in front of the fraction remains, as I removed the parentheses by multiplying by a positive 5.

$$2\left(3 - \frac{5x - 5}{2} = x\right)$$

$$6 - \frac{2(5x - 5)}{2} = 2x$$

$$6 - 5x + 5 = 2x$$

$$11 - 5x = 2x$$

$$11 = 2x + 5x$$

$$11 = 7x$$

$$\frac{11}{7} = \frac{7x}{7}$$

$$\frac{11}{7} = x$$

$$x = \frac{11}{7}$$

9. $x = -\frac{1}{3}$

10. $1 - \dfrac{x}{2} = 5$

$$2\left(1 - \dfrac{x}{2} = 5\right)$$

$$2 - x = 10$$
$$-x = 10 - 2$$
$$-x = 8$$
$$x = -8$$

UNIT 4

1. $x = 10$ The equation is conditional.

2. $1 - 2[4 - (3x - 5)] = 3x + 1$
    $1 - 2[4 - 3x + 5] = 3x + 1$
    $1 - 8 + 6x - 10 = 3x + 1$
    $6x - 17 = 3x + 1$
    $6x - 3x = 1 + 17$
    $3x = 18$
    $\dfrac{3x}{3} = \dfrac{18}{3}$
    $x = 6$ The equation is conditional.

3. $x = 1$ The equation is conditional.

4. $2\{x + 4 - 3(2x - 1)\} = 3(4 - 3x) + 2 - x$
    $2\{x + 4 - 6x + 3\} = 12 - 9x + 2 - x$
    $2x + 8 - 12x + 6 = -10x + 14$
    $-10x + 14 = -10x + 14$
    An identity, so the solution is the set of reals.

5. $x = 0$ The equation is conditional.

6. $12x + 4 - [(3 - x) - (5x + 7)] = 13$

$12x + 4 - [3 - x - 5x + 7] = 13$

$12x + 4 - 3 + x + 5x + 7 = 13$

$18x + 8 = 13$

$18x = 13 - 8$

$18x = 5$

$\dfrac{18x}{18} = \dfrac{5}{18}$

$x = \dfrac{5}{18}$ The equation is conditional.

7. $x = -1$ The equation is conditional.

8. $x - \dfrac{x+2}{2} = 3x - (1 + 2x) - \dfrac{x}{2}$

$x - \dfrac{x+2}{2} = 3x - 1 - 2x - \dfrac{x}{2}$

$2\left(x - \dfrac{x+2}{2} = x - 1 - \dfrac{x}{2} \right)$

$2x - \dfrac{2(x+2)}{2} = 2x - 2 - \dfrac{2x}{2}$

$2x - x - 2 = 2x - 2 - x$

$x - 2 = x - 2$

An identity, and the solution is all real numbers.

9. An identity, and the solution is all real numbers.

10. $20 - \{[3x - (x - 1)] - 5x\} = 0$

$20 - \{[3x - x + 1] - 5x\} = 0$

$20 - \{3x - x + 1 - 5x\} = 0$

$20 - 3x + x - 1 + 5x = 0$

$19 + 3x = 0$

$3x = -19$

$\dfrac{3x}{3} = \dfrac{-19}{3}$

$x = -\dfrac{19}{3}$ The equation is conditional.

UNIT 5

1. $x = 5$

2. $$\frac{x-3}{2} = \frac{2x+4}{5}$$

    $5(x-3) = 2(2x+4)$ Cross-multiply.

    $5x - 15 = 4x + 8$

    $5x - 4x = 8 + 15$

    $\quad\quad x = 23$

    It is not necessary to check this, as we did not have a fractional equation—the variable was not in any denominator.

3. $x = 0$

4. $$\frac{3x-3}{x-1} = 2$$

    $3x - 3 = 2(x-1)$ Cross-multiply.

    $3x - 3 = 2x - 2$

    $3x - 2x = -2 + 3$

    $\quad\quad x = 1$

    Check: $\dfrac{3-3}{1-1} \overset{?}{=} 2$

    But we cannot have 0 in the denominator; therefore there is no solution to this equation.

5. $x = 4$

6. $$\frac{3}{x} = \frac{4}{x-2}$$

    $3(x-2) = 4x$

    $3x - 6 = 4x$

    $3x - 4x = 6$

    $\quad -x = 6$

    $\quad\quad x = -6$

    Check: $\dfrac{3}{x} = \dfrac{4}{x-2}$

    $\dfrac{3}{(-6)} \overset{?}{=} \dfrac{4}{(-6)-2}$

    $-\dfrac{1}{2} = -\dfrac{1}{2}$

    True; therefore the solution is $x = -6$.

7. $x = 5$

8. $\dfrac{5-2x}{x-1} = -2$

$$5 - 2x = -2(x-1)$$
$$5 - 2x = -2x + 2$$
$$-2x + 2x = 2 - 5$$
$$0 = -3$$

No solution, since 0 does not equal –3.

9. $x = 7$

10. $\dfrac{2}{x} + \dfrac{3(x-1)}{5x} = 1$

$$\left[\dfrac{2}{x} + \dfrac{3(x-1)}{5x} = 1\right]5x$$

$$\dfrac{2}{x} \cdot 5x + \dfrac{3(x-1)}{5x} \cdot 5x = 1 \cdot 5x$$

$$10 + 3(x-1) = 5x$$
$$10 + 3x - 3 = 5x$$
$$3x + 7 = 5x$$
$$3x - 5x = -7$$
$$-2x = -7$$
$$\dfrac{-2x}{-2} = \dfrac{-7}{-2}$$
$$x = \dfrac{7}{2}$$

You may have the pleasure of checking to prove that $x = \frac{7}{2}$ is the solution.

11. $x = \frac{1}{3}$

12. $\dfrac{4}{x-2} - \dfrac{1}{x} = \dfrac{5}{x-2}$

$$\left(\dfrac{4}{x-2} - \dfrac{1}{x} = \dfrac{5}{x-2}\right)(x-2)$$

$$\dfrac{4}{x-2} \cdot (x-2) - \dfrac{1 \cdot (x-2)}{x} = \dfrac{5}{x-2} \cdot (x-2)$$

$$4 - \dfrac{x-2}{x} = 5$$

$$4 \cdot x - \dfrac{(x-2) \cdot x}{x} = 5 \cdot x$$

$$4x - x + 2 = 5x$$
$$2 = 5x + x - 4x$$
$$2 = 2x$$
$$\dfrac{2}{2} = \dfrac{2x}{2}$$
$$1 = x$$

UNIT 6

1. $a = \dfrac{3cy}{2mx}$

2. $\qquad 2cy + 4d = 3ax - 4b$

    $2cy + 4d + 4b = 3ax$

    $\dfrac{2cy + 4d + 4b}{3x} = \dfrac{\cancel{3ax}}{\cancel{3x}}$

    $\dfrac{2cy + 4d + 4b}{3x} = a$

3. $a = \dfrac{bx + 7c}{x + 3}$

4. $3(x - a) = 2a - x - \dfrac{b - x}{c}$

    $\left(3x - 3a = 2a - x - \dfrac{b - x}{c} \right) c$

    $3cx - 3ac = 2ac - cx - (b - x) \leftarrow$ Note parentheses.

    $3cx - 3ac = 2ac - cx - b + x$

    $-3ac - 2ac = -cx - b + x - 3cx$

    $-5ac = -4cx - b + x$

    $\dfrac{\cancel{-5ac}}{\cancel{-5c}} = \dfrac{-4cx - b + x}{-5c}$

    $a = \dfrac{-4cx - b + x}{-5c}$

    Or, if you prefer:

    $a = \dfrac{4cx + b - x}{5c}$

5. $a = \dfrac{\pi - cy}{x + 2}$

6. $\dfrac{2ax}{3c} = \dfrac{y}{m}$

    $2amx = 3cy$

    $\dfrac{\cancel{2amx}}{\cancel{2am}} = \dfrac{3cy}{2am}$

    $x = \dfrac{3cy}{2am}$

7. $\dfrac{2cy + 4d + 4b}{3a} = x$

8. $ax + 3a = bx + 7c$

$ax - bx = 7c - 3a$

$x(a - b) = 7c - 3a$

$\dfrac{x\cancel{(a-b)}}{\cancel{a-b}} = \dfrac{7c - 3a}{a - b}$

$x = \dfrac{7c - 3a}{a - b}$

9. $x = \dfrac{5ac - b}{4c - 1}$

10. $a(x + 2) = \pi - cy$

$ax + 2a = \pi - cy$

$ax = \pi - cy - 2a$

$\dfrac{\cancel{a}x}{\cancel{a}} = \dfrac{\pi - cy - 2a}{a}$

$x = \dfrac{\pi - cy - 2a}{a}$

11. $c = \dfrac{2amx}{3y}$

12. $2cy + 4d = 3ax - 4b$

$2cy = 3ax - 4b - 4d$

$\dfrac{\cancel{2cy}}{\cancel{2y}} = \dfrac{3ax - 4b - 4d}{2y}$

$c = \dfrac{3ax - 4b - 4d}{2y}$

13. $c = \dfrac{ax + 3a - bx}{7}$

14. $3(x - a) = 2a - x - \dfrac{b - x}{c}$

$\left(3(x - a) = 2a - x - \dfrac{b - x}{c}\right)c$

$3(x - a)c = 2ac - cx - \dfrac{b - x}{\cancel{c}} \cdot \cancel{c}$

$3cx - 3ac = 2ac - cx - b + x$

$3cx - 3ac - 2ac + cx = x - b$

$4cx - 5ac = x - b$

$c(4x - 5a) = x - b$

$\dfrac{c\cancel{(4x - 5a)}}{\cancel{4x - 5a}} = \dfrac{x - b}{4x - 5a}$

$c = \dfrac{x - b}{4x - 5a}$

15. $c = \dfrac{\pi - ax - 2a}{y}$

UNIT 7

1. $x - 5$

2. $3x + 8$

3. $8x - 10$

4. $\dfrac{x}{3}$

5. $2x - 5 = 11$

6. $7x = 35$

7. $x + 20 = 32$

8. $x + 12 = 20$

9. $15 + 2x = 47$

10. $4 + 3x = 17$

11. Jack is 17 and George is 25.

12.   Let x = length of shorter piece of rope.
    Then $x + 10$ = length of longer piece.
    The length of the rope is 36 feet.

$$x + (x + 10) = 36$$
$$2x = 26$$
$$x = 13$$
$$x + 10 = 23$$

The shorter piece of rope is 13 feet long, and the other is 23 feet long.

UNIT 8

1. $(3y)^2 \cdot (2y)^3 = 3^2y^2 \cdot 2^3y^3 = 9y^2 \cdot 8y^3 = 72y^5$

2. $3x^0 = 3 \cdot 1 = 3$

3. $x^2(x^3)^4 = x^2 \cdot x^{12} = x^{2+12} = x^{14}$

4. $\left(\dfrac{a^2b^3cd^5}{3x^2w^0}\right)^7 = \dfrac{a^{14}b^{21}c^7d^{35}}{3^7x^{14}}$

5. $\dfrac{(2ab)^2}{(3x^3)^2} = \dfrac{2^2a^2b^2}{3^2x^6} = \dfrac{4a^2b^2}{9x^6}$

6. $(3x^5)^2(2x^3)^3 = 3^2x^{10} \cdot 2^3x^9 = 9x^{10} \cdot 8x^9 = 72x^{19}$

7. $(x^2y)(xy^2) = x^{2+1} \cdot y^{1+2} = x^3y^3$

8. $2(3ab^2)^2 = 2 \cdot 3^2a^2b^4 = 18a^2b^4$

9. $(-4c)^2 = (-4)^2c^2 = 16c^2$

10. $\left(\dfrac{xyz^2}{5a}\right)^3 = \dfrac{x^3y^3z^6}{5^3a^3} = \dfrac{x^3y^3z^6}{125a^3}$

11. $(-2abc)(bcd)(3abc^2) = -2 \cdot 3a^{1+1} \cdot b^{1+1+1} \cdot c^{1+1+2} \cdot d = -6a^2b^3c^4d$

12. $(-2x^2yz)(-5xz)^2(xyz^2)^3 = 2x^2yz \cdot (-5)^2x^2z^2 \cdot x^3y^3z^6$

    $= 2 \cdot 25x^{2+2+3} \cdot y^{1+3} \cdot z^{1+2+6}$

    $= 50x^7y^4z^9$

UNIT 9

1. $\dfrac{a^{-3}}{a^2} = \dfrac{1}{a^3a^2} = \dfrac{1}{a^5}$

2. $\dfrac{a^{-2}x^3}{y^{-1}} = \dfrac{x^3y}{a^2}$

3. $(x^2y)^{-2} = x^{-4}y^{-2} = \dfrac{1}{x^4y^2}$

4. $\dfrac{(ab^2)^{-3}}{(x^2y^{-3})^4} = \dfrac{a^{-3}b^{-6}}{x^8y^{-12}} = \dfrac{y^{12}}{a^3b^6x^8}$

5. $\dfrac{(3ab^5)^{-3}}{2x^{-5}} = \dfrac{3^{-3}a^{-3}b^{-15}}{2x^{-5}} = \dfrac{x^5}{2 \cdot 3^3a^3b^{15}} = \dfrac{x^5}{54a^3b^{15}}$

6. $\dfrac{7x^{-1}}{y^2} = \dfrac{7}{xy^2}$

7. $(5w^{-2})^2(2w^{-2}) = 5^2w^{-4} \cdot 2w^{-2} = \dfrac{5^2 \cdot 2}{w^4 \cdot w^2} = \dfrac{50}{w^6}$

8. $\dfrac{x^{-2}y^{-3}}{(c)^{-2}} = \dfrac{x^{-2}y^{-3}}{c^{-2}} = \dfrac{c^2}{x^2y^3}$

9. $\dfrac{(5a^2b^3)^2}{(-2x)^{-3}} = \dfrac{5^2a^4b^6}{(-2)^{-3}x^{-3}} = (-2)^3 5^2 a^4b^6x^3 = 200a^4b^6x^3$

10. $\dfrac{16w^{-1}y^2z^{-3}}{2x} = \dfrac{16y^2}{2wxz^3} = \dfrac{8y^2}{wxz^3}$

UNIT 10

1. $\dfrac{x^{-4}}{x^4} = \dfrac{1}{x^8}$

2. $\dfrac{15x^5 y^3}{3x^2 y^7} = \dfrac{15\cancel{x^5}\cancel{y^3}}{3\cancel{x^2}\cancel{y^7}} = \dfrac{5x^{5-2}}{y^{7-3}} = \dfrac{5x^3}{y^4}$

3. $\dfrac{x^5 \cdot x^{-4}}{x^{-3}} = x^4$

4. $x(3x^2 y^{-3})^2 = x \cdot 3^2 x^4 y^{-6} = \dfrac{9x \cdot x^4}{y^6} = \dfrac{9x^5}{y^6}$

5. $(2w^{-2})^2 (5w^{-2}) = \dfrac{20}{w^6}$

6. $x(5xy^{-2})^{-2} = x \cdot 5^{-2} x^{-2} y^{+4} = \dfrac{\cancel{x}\, y^4}{5^2 \cancel{x^2}} = \dfrac{y^4}{25x^{2-1}} = \dfrac{y^4}{25x}$

7. $\dfrac{7a^{-4} b^4}{28a^{-3} b^{-3}} = \dfrac{b^7}{4a}$

8. $\dfrac{(3x^2 y)^{-1}}{2xy^{-5}} = \dfrac{3^{-1} x^{-2} y^{-1}}{2xy^{-5}} = \dfrac{y^5}{2x \cdot 3x^2 y} = \dfrac{y^4}{6x^3}$

9. $\dfrac{(m^{-3} s^{-3})^3}{m^{-4} s^4} = \dfrac{1}{m^5 s^{13}}$

10. $\dfrac{3x^{-2} y^3 c^5}{x^{-3} y^7 c^2} = \dfrac{3\cancel{x^3}\cancel{y^3}\cancel{c^5}}{\cancel{x^2}\cancel{y^7}\cancel{c^2}} = \dfrac{3x^{3-2} c^{5-2}}{y^{7-3}} = \dfrac{3xc^3}{y^4}$

11. $\dfrac{(3xy^{-2})^{-3}}{x} = \dfrac{y^6}{27x^4}$

12. $\left(\dfrac{a^{-1} b^2}{a^{-3} b^{-2}}\right)^{-2} = \dfrac{a^2 b^{-4}}{a^6 b^4} = \dfrac{\cancel{a^2}}{\cancel{a^6} b^4 b^4} = \dfrac{1}{a^{6-2} b^{4+4}} = \dfrac{1}{a^4 b^8}$

13. $\left[\dfrac{x^{-4} y^{-3} z^2}{x^{-3} y^2 z^{-4}}\right]^{-2} = \dfrac{x^2 y^{10}}{z^{12}}$

14. $\left[\dfrac{(ab)^{-1}}{(a^{-2} b^3)^3}\right]^{-1} = \left[\dfrac{a^{-1} b^{-1}}{a^{-6} b^9}\right]^{-1} = \dfrac{ab}{a^6 b^{-9}} = \dfrac{abb^9}{a^6} = \dfrac{b^{1+9}}{a^{6-1}} = \dfrac{b^{10}}{a^5}$

UNIT 11

1. $\dfrac{3}{11}$

2. $\dfrac{7}{10} - \dfrac{9}{10} = \dfrac{7-9}{10} = \dfrac{-2}{10} = \dfrac{-1}{5}$

3. $\dfrac{44}{45}$

4. $\dfrac{2}{13} \diagdown \dfrac{6}{-5} = \dfrac{-10+78}{-65} = \dfrac{68}{-65} = -\dfrac{68}{65}$

5. $\dfrac{-1}{4}$

6. $\dfrac{-6}{-15} \diagdown \dfrac{3}{5} = \dfrac{-30-45}{-75} = \dfrac{-75}{-75} = 1$

7. $\dfrac{29}{90}$

8. $\dfrac{11}{t} \diagdown \dfrac{7}{r} = \dfrac{11r+7t}{rt}$

9. $\dfrac{2}{x}$

10. $\dfrac{10}{x+1} + \dfrac{3}{x+1} = \dfrac{10+3}{x+1} = \dfrac{13}{x+1}$

11. $\dfrac{1}{a}$

12. $\dfrac{-s}{9} \diagdown \dfrac{k}{10} = \dfrac{-10s+9k}{90}$

13. $\dfrac{3x}{10}$

14. $\dfrac{x+1}{2} \diagdown \dfrac{3}{5} = \dfrac{5(x+1)-6}{10}$

$= \dfrac{5x+5-6}{10}$

$= \dfrac{5x-1}{10}$

15. $\dfrac{-x-5}{6}$

16. $\dfrac{x+2}{2} \diagdown \dfrac{x+3}{3} = \dfrac{3(x+2)-2(x+3)}{6}$

$= \dfrac{3x+6-2x-6}{6}$

$= \dfrac{x}{6}$

UNIT 12

1. $\dfrac{5}{3}$

2. $\dfrac{5}{18} \div \dfrac{3}{14} = \dfrac{5}{\overset{}{\underset{9}{\cancel{18}}}} \cdot \dfrac{\overset{7}{\cancel{14}}}{3} = \dfrac{35}{27}$

3. $\dfrac{-2}{3}$

4. $\dfrac{9}{14} \div \dfrac{5}{21} = \dfrac{9}{\underset{2}{\cancel{14}}} \cdot \dfrac{\overset{3}{\cancel{21}}}{5} = \dfrac{27}{10}$

5. $\dfrac{14}{27}$

6. $\left(\dfrac{2}{3} \cdot \dfrac{4}{5}\right) \cdot \dfrac{10}{10} = \left(\dfrac{8}{15}\right) \cdot \dfrac{10}{10} = \dfrac{8}{15} \cdot \dfrac{1}{1} = \dfrac{8}{15}$

7. 3

8. $\dfrac{3-\frac{2}{5}}{3+\frac{2}{5}} = \left(\dfrac{3}{1}-\dfrac{2}{5}\right) \div \left(\dfrac{3}{1}+\dfrac{2}{5}\right)$

$= \dfrac{15-2}{5} \div \dfrac{15+2}{5}$

$= \dfrac{13}{5} \div \dfrac{17}{5}$

$= \dfrac{13}{\cancel{5}} \cdot \dfrac{\cancel{5}}{17}$

$= \dfrac{13}{17}$

9. $\dfrac{4}{5}$

10. $\dfrac{\frac{a}{2}-\frac{3}{5}}{2} = \left(\dfrac{a}{2}-\dfrac{3}{5}\right) \div 2$

$= \dfrac{5a-6}{10} \div \dfrac{2}{1}$

$= \dfrac{5a-6}{10} \cdot \dfrac{1}{2}$

$= \dfrac{5a-6}{20}$

11. $\dfrac{28}{x-1}$

12. $\dfrac{\frac{2}{x}-5}{x} = \left(\dfrac{2}{x}-\dfrac{5}{1}\right) \div x$

$= \dfrac{2-5x}{x} \cdot \dfrac{1}{x}$

$= \dfrac{2-5x}{x^2}$

13. $\dfrac{36-x}{2}$

14. $\dfrac{\frac{x}{5}-\frac{x+2}{2}}{\frac{x}{2}} = \left(\dfrac{x}{5}-\dfrac{x+2}{2}\right) \div \dfrac{x}{2}$

$= \dfrac{2x-5(x+2)}{10} \cdot \dfrac{2}{x}$

$= \dfrac{2x-5x-10}{10} \cdot \dfrac{2}{x}$

$= \dfrac{-3x-10}{\underset{5}{\cancel{10}}} \cdot \dfrac{\cancel{2}}{x}$

$= \dfrac{-3x-10}{5x}$

15. $\dfrac{a+2b}{a+b}$

16. $\dfrac{4}{\frac{3}{2}-\frac{x+1}{x}} = 4 \div \left(\dfrac{3}{2}-\dfrac{x+1}{x}\right)$

$= 4 \div \left[\dfrac{3x-2(x+1)}{2x}\right]$

$= 4 \div \dfrac{3x-2x-2}{2x}$

$= 4 \div \dfrac{x-2}{2x}$

$= \dfrac{4}{1} \cdot \dfrac{2x}{x-2}$

$= \dfrac{8x}{x-2}$

UNIT 13

1. 2

2. $(-1)^{2/3} = (\sqrt[3]{-1})^2 = (-1)^2 = 1$

3. $\sqrt{-4}$
   No solution; we cannot take the square root of a negative number in the set of reals.

4. $4^{3/2} = (\sqrt{4})^3 = (2)^3 = 8$

5. 2

6. $4^{-1/2} = (\sqrt{4})^{-1} = 2^{-1} = \dfrac{1}{2}$

7. $\dfrac{1}{\sqrt{x}}$

8. $x^{1/3} = \sqrt[3]{x}$

9. $\sqrt[5]{a^2}$

10. $4^{-3/2} = (\sqrt{4})^{-3} = 2^{-3} = \dfrac{1}{2^3} = \dfrac{1}{8}$

11. $\sqrt{x+1}$

12. $x^{8/3} = \sqrt[3]{x^8} = \sqrt[3]{x^3 \cdot x^3 \cdot x^2} = x \cdot x\sqrt[3]{x^2} = x^2\sqrt[3]{x^2}$

13. $2\sqrt{x}$

14. $x^{11/2} = \sqrt{x^{11}} = \sqrt{x^{10} \cdot x} = x^5\sqrt{x}$

15. $\dfrac{1}{\sqrt{5x}}$

16. $\sqrt{18x^3} = \sqrt{9 \cdot 2 \cdot x^2 x} = \sqrt{9x^2 \cdot 2x} = 3x\sqrt{2x}$

17. $\sqrt[3]{4x^2}$

18. $(-64)^{2/3} = (\sqrt[3]{-64})^2 = (-4)^2 = 16$

19. $(7x)^{1/2}$

20. $\sqrt[3]{2x} = (2x)^{1/3}$

UNIT 14

1. $\dfrac{y^{2/3}}{y^{1/3}} = y^{2/3\,-1/3} = y^{1/3} = \sqrt[3]{y}$

2. $(y^{3/5})^{1/4} = y^{3/5\,\cdot 1/4} = y^{3/20} = \sqrt[20]{y^3}$

3. $x^{1/2} \cdot x^{3/5} = x^{1/2\,+3/5} = x^{(5+6)/10} = x^{11/10} = \sqrt[10]{x^{11}} = \sqrt[10]{x^{10} \cdot x} = x\sqrt[10]{x}$

4. $\left(\dfrac{a^4}{c^2}\right)^{1/2} = \dfrac{a^2}{c}$

5. $[(3\sqrt{4})^{-1}]^2 = [(3 \cdot 2)^{-1}]^2$

    $\qquad\qquad\;\; = [6^{-1}]^2$

    $\qquad\qquad\;\; = 6^{-2}$

    $\qquad\qquad\;\; = \dfrac{1}{6^2}$

    $\qquad\qquad\;\; = \dfrac{1}{36}$

6. $(x\sqrt{x})^{1/2} = (x \cdot x^{1/2})^{1/2}$

    $\qquad\qquad = x^{1/2} \cdot x^{1/4}$

    $\qquad\qquad = x^{1/2\,+1/4}$

    $\qquad\qquad = x^{(4+2)/8}$

    $\qquad\qquad = x^{6/8}$

    $\qquad\qquad = x^{3/4}$

    $\qquad\qquad = \sqrt[4]{x^3}$

7. $(8x^2)^{1/3} = 8^{1/3}\,x^{2/3} = 2\sqrt[3]{x^2}$

8.
$$\left(\frac{2^{-3} \cdot 2^5}{2^{-2}}\right)^3 = \frac{2^{-9} \cdot 2^{15}}{2^{-6}}$$
$$= \frac{2^{15} \cdot 2^6}{2^9}$$
$$= \frac{2^{21}}{2^9}$$
$$= 2^{21-9}$$
$$= 2^{12}$$

An alternative, shorter approach would be

$$\left(\frac{2^{-3} \cdot 2^5}{2^{-2}}\right)^3 = \left(\frac{2^2}{2^{-2}}\right)^3$$
$$= (2^2 \cdot 2^2)^3$$
$$= (2^4)^3$$
$$= 2^{12}$$

9. $\left(\dfrac{x^{1/3}}{x^{2/3}}\right)^3 = \dfrac{x}{x^2} = \dfrac{1}{x}$

10. $x^{1/2} \cdot x^{5/2} = x^{1/2 + 5/2} = x^{6/2} = x^3$

11. $\left(\sqrt[3]{x^2}\right)^{1/2} = (x^{2/3})^{1/2} = x^{1/3} = \sqrt[3]{x}$

12.
$$\frac{x^{-7/2} \cdot x^{3/2}}{\sqrt{x} \cdot x^{-3/2}} = \frac{x^{-7/2} \cdot x^{3/2}}{x^{1/2} \cdot x^{-3/2}}$$
$$= \frac{x^{3/2} \cdot x^{3/2}}{x^{7/2} \cdot x^{1/2}}$$
$$= \frac{x^{3/2 + 3/2}}{x^{7/2 + 1/2}}$$
$$= \frac{x^{6/2}}{x^{8/2}}$$
$$= \frac{x^3}{x^4}$$
$$= \frac{1}{x}$$

13.
$$(8\sqrt{x})^{-2/3} = (8x^{1/2})^{-2/3}$$
$$= 8^{-2/3} x^{-1/3}$$
$$= \frac{1}{8^{+2/3} x^{1/3}}$$
$$= \frac{1}{(\sqrt[3]{8})^2 \sqrt[3]{x}}$$
$$= \frac{1}{4\sqrt[3]{x}}$$

14.
$$\left(\frac{27^{5/3} \cdot 27^{-1/3}}{27^{1/3}}\right)^2 = \frac{27^{10/3} \cdot 27^{-2/3}}{27^{2/3}}$$
$$= \frac{27^{10/3}}{27^{2/3} \cdot 27^{2/3}}$$
$$= \frac{27^{10/3}}{27^{2/3 + 2/3}}$$
$$= \frac{27^{10/3}}{27^{4/3}}$$
$$= 27^{10/3 - 4/3}$$
$$= 27^{6/3}$$
$$= 27^2$$

15.
$$\left(\frac{3x^{-1}}{\sqrt{x}}\right)^2 = \left(\frac{3x^{-1}}{x^{1/2}}\right)^2$$
$$= \frac{3^2 x^{-2}}{x}$$
$$= \frac{9}{x \cdot x^2}$$
$$= \frac{9}{x^3}$$

UNIT 15

1. $\dfrac{1}{(x^2+1)^2}$

2. $x^{-1}+2^{-2}=\dfrac{1}{x}+\dfrac{1}{2^2}$

$$=\dfrac{1}{x}+\dfrac{1}{4}$$

$$=\dfrac{4+x}{4x}$$

3. $\dfrac{x}{1-x}$

4. $\dfrac{1}{x^{-1}+y^{-1}}=\dfrac{1}{\dfrac{1}{x}+\dfrac{1}{y}}$

$$=1\div\left(\dfrac{1}{x}+\dfrac{1}{y}\right)$$

$$=1\div\left(\dfrac{y+x}{xy}\right)$$

$$=1\cdot\dfrac{xy}{y+x}$$

$$=\dfrac{xy}{y+x}$$

5. $\dfrac{4}{9}$

6. $\dfrac{x^{-1}+2y^{-1}}{3}=\dfrac{\dfrac{1}{x}+\dfrac{2}{y}}{3}$ $\leftarrow$ Be careful that the 2 is in the numerator.

$$=\left(\dfrac{1}{x}+\dfrac{2}{y}\right)\div3$$

$$=\dfrac{y+2x}{xy}\div3$$

$$=\dfrac{y+2x}{xy}\cdot\dfrac{1}{3}$$

$$=\dfrac{y+2x}{3xy}$$

7. $\dfrac{ab+1}{ab}$

8. $5(x+y)^{-1}=\dfrac{5}{(x+y)}$

9. $\dfrac{(2x-3)^2}{x}$

10. $3x^{-2}+y=\dfrac{3}{x^2}+y$

$$=\dfrac{3}{x^2}+\dfrac{y}{1}$$

$$=\dfrac{3+x^2y}{x^2}$$

11. $\dfrac{a^2-1}{a^2+1}$

12. $\dfrac{3^{-1}+2^{-1}}{3^{-1}-2^{-1}} = \dfrac{\frac{1}{3}+\frac{1}{2}}{\frac{1}{3}-\frac{1}{2}}$

$$= \left(\frac{1}{3}+\frac{1}{2}\right) \div \left(\frac{1}{3}-\frac{1}{2}\right)$$

$$= \frac{2+3}{6} \div \frac{2-3}{6}$$

$$= \frac{5}{6} \div \frac{-1}{6}$$

$$= \frac{5}{\cancel{6}} \cdot \frac{\cancel{6}}{-1}$$

$$= -5$$

13. $a(ab+1)$

14. $\dfrac{3ab^{-1}}{a^{-1}+b} = \dfrac{\frac{3a}{b}}{\frac{1}{a}+b}$

$$= \frac{3a}{b} \div \left(\frac{1}{a}+b\right)$$

$$= \frac{3a}{b} \div \left(\frac{1+ab}{a}\right)$$

$$= \frac{3a}{b} \cdot \frac{a}{1+ab}$$

$$= \frac{3a^2}{b(1+ab)}$$

UNIT 16

1. $10c^3x^2 - 2c^2x^2 - 6cx^3$

2. $(x+4)(x+5) = x^2 + 9x + 20$

3. $x^2 - 9x + 14$

4. $(x-1)(x-5) = x^2 - 6x + 5$

5. $x^2 - x - 6$

6. $(a+5)^2 = a^2 + 2(5a) + 25$
   $\qquad\quad = a^2 + 10a + 25$

7. $x^2 - 4$

8. $(x-1)^2 = x^2 - 2x + 1$

9. $2x^2 + x - 3$

10. $(5x-2y)(3x+7y) = 5x(3x+7y) - 2y(3x+7y)$
    $\qquad\qquad\qquad\quad = 15x^2 + 35xy - 6xy - 14y^2$
    $\qquad\qquad\qquad\quad = 15x^2 + 29xy - 14y^2$

11. $6x^2 - 13x - 5$

12. $x(x-4)^2 = x(x^2 - 8x + 16)$
    $\qquad\qquad = x^3 - 8x^2 + 16x$

13. $x^3 - 10x^2 + 25x$

14. $(x - 2)(x^3 - 4x^2 + 7x - 1) = x(x^3 - 4x^2 + 7x - 1) - 2(x^3 - 4x^2 + 7x - 1)$
$$= x^4 - 4x^3 + 7x^2 - x - 2x^3 + 8x^2 - 14x + 2$$
$$= x^4 - 6x^3 + 15x^2 - 15x + 2$$

15. $x^4 - 2x^2 - 3$

16. $(x + 2y)(x - 3y) = x(x - 3y) + 2y(x - 3y)$
$$= x^2 - 3xy + 2xy - 6y^2$$
$$= x^2 - xy - 6y^2$$

17. $-2a^4 + a^3 - 2a^2 + 7a - 3$

18. $(x^2 - 3x + 1)(x^3 - 2x) = x^2(x^3 - 2x) - 3x(x^3 - 2x) + 1(x^3 - 2x)$
$$= x^5 - 2x^3 - 3x^4 + 6x^2 + x^3 - 2x$$
$$= x^5 - 3x^4 - x^3 + 6x^2 - 2x$$

19. $x^3 - 3x^2 + 5x - 15$

20. $2xy^2(x + y)(x - 3) = 2xy^2[x(x - 3) + y(x - 3)]$
$$= 2xy^2[x^2 - 3x + xy - 3y]$$
$$= 2x^3y^2 - 6x^2y^2 + 2x^2y^3 - 6xy^3$$

21. $5x^2 + 8x - 4$

22. $-3(x - 1)(x - 2) = -3(x^2 - 3x + 2)$
$$= -3x^2 + 9x - 6$$

23. $6a^2 - 4ab - 2b^2$

24. $(5a - 3b)(-2a + 6b) = 5a(-2a + 6b) - 3b(-2a + 6b)$
$$= -10a^2 + 30ab + 6ab - 18b^2$$
$$= -10a^2 + 36ab - 18b^2$$

UNIT 17

1.
$$\begin{array}{r}
x + 3 + \dfrac{1}{x + 5} \\
x + 5 \overline{)\, x^2 + 8x + 16} \\
\underline{x^2 + 5x} \\
3x + 16 \\
\underline{3x + 15} \\
1
\end{array}$$

2.
$$\begin{array}{r}
2x - 3 + \dfrac{4}{5x + 1} \\
5x + 1 \overline{)\, 10x^2 - 13x + 1} \\
\underline{10x^2 + 2x} \\
-15x + 1 \\
\underline{-15x - 3} \\
4
\end{array}$$

3.

$$\begin{array}{r} 3x^3 + 5x^2 + 5x + 12 + \dfrac{2}{x-1} \\[4pt] x-1 \overline{\smash{\big)}\ 3x^4 + 2x^3 + 0 + 7x - 10} \\ \underline{3x^4 - 3x^3} \\ 5x^3 + 0 \\ \underline{5x^3 - 5x^2} \\ 5x^2 + 7x \\ \underline{5x^2 - 5x} \\ 12x - 10 \\ \underline{12x - 12} \\ 2 \end{array}$$

4.

$$\begin{array}{r} 3x \;+\; \dfrac{-8x + 2}{2x^2 + 3x - 1} \\[4pt] 2x^2 + 3x - 1 \overline{\smash{\big)}\ 6x^3 + 9x^2 - 11x + 2} \\ \underline{6x^3 + 9x^2 - 3x} \\ -8x + 2 \end{array}$$

5.

$$\begin{array}{r} x^2 + 3x \;+ 5 + \dfrac{13}{2x - 3} \\[4pt] 2x-3 \overline{\smash{\big)}\ 2x^3 + 3x^2 + x - 2} \\ \underline{2x^3 - 3x^2} \\ 6x^2 + x \\ \underline{6x^2 - 9x} \\ 10x - 2 \\ \underline{10x - 15} \\ 13 \end{array}$$

6.

$$\begin{array}{r} x^2 - 2x + 4 + \dfrac{-6}{x + 2} \\[4pt] x+2 \overline{\smash{\big)}\ x^3 + 0 \;\; + 0 + 2} \\ \underline{x^3 + 2x^2} \\ -2x^2 + 0 \\ \underline{-2x^2 - 4x} \\ 4x + 2 \\ \underline{4x + 8} \\ -6 \end{array}$$

7.

$$\begin{array}{r} 3x \;- 1 \;\; + \dfrac{-x + 6}{x^2 + x + 1} \\[4pt] x^2+x+1 \overline{\smash{\big)}\ 3x^3 + 2x^2 + \; x + 5} \\ \underline{3x^3 + 3x^2 + 3x} \\ -x^2 - 2x + 5 \\ \underline{-x^2 - \; x - 1} \\ - \; x + 6 \end{array}$$

UNIT 18

1. $(x + 3)(x - 1)$

2. $(x - 8)(x - 7)$

3. $x(x + 1)$

4. $3xy(x - 4y)$

5. $3b(x^2 + 9b)$

6. $(x - 3)(x + 2)$

7. $(x + 3)(x + 2)$

8. $(x - 3)(x - 4)$

9. Prime

10. $(x + 2)(x - 4)$

11. Prime

12. $2x(x^2 + x + 11)$

13. $5(x^2 - x - 1)$

14. $(x + 5)(x - 2)$

15. $(x + 10)(x - 3)$

16. $(x + 6)(x + 1)$

17. $(x + 1)(x + 1)$

18. $(x - 3)(x - 3)$

19. $(x - 8)(x + 7)$

20. $(x + 9)(x - 5)$

21. $(x + 8)(x + 8)$

22. $(x - 5)(x - 8)$

23. $(x + 9)(x - 2)$

24. $x(x^2 + x + 5)$

25. $(x + 7)(x + 3)$

26. $(x - 9)(x + 2)$

UNIT 19

1. $5c(c - 1)$

2. $m^2 + 10m + 25 = (m + 5)^2$

3. $(x - 10)(x + 1)$

4. $2 - 2x^2 = 2(1 - x^2)$
   $= 2(1 + x)(1 - x)$

5. $3(b + 5)(b - 5)$

6. $9 - 6x + x^2 = x^2 - 6x + 9$
   $= (x - 3)(x - 3)$

7. $3ab(c - d)(c + d)$

8. $8x^3 + 27y^3 = (2x + 3y)(4x^2 - 6xy + 9y^2)$

9. $(x + 4)(x - 2)$

10. $x^2 - x + 7$ prime

11. $x(x + 6)(x - 6)$

12. $x^7 - 8x^4y^3 = x^4(x^3 - 8y^3)$
    $$= x^4(x - 2y)(x^2 + 2xy + 4y^2)$$

13. $(x + 10)(x + 3)$

14. $3r^3 - 6r^2 - 45r = 3r(r^2 - 2r - 15)$
    $$= 3r(r - 5)(r + 3)$$

15. $(x + 7)(x - 2)$

16. $2a^2b^2c^2 - 4ab^2c^2 + 2b^2c^2 = 2b^2c^2(a^2 - 2a + 1)$
    $$= 2b^2c^2(a - 1)(a - 1)$$

17. $5y(x^2 - 3x - 2)$

18. $5x^4 + 10x^3 - 15x^2 = 5x^2(x^2 + 2x - 3)$
    $$= 5x^2(x + 3)(x - 1)$$

19. $3(x - 2)(x + 2)$

20. $a^2b^2 - a^2c^2 = a^2(b^2 - c^2)$
    $$= a^2(b + c)(b - c)$$

21. $2x(y^2 - 27y + 50)$

22. $10ab^2 - 140ab + 330a = 10a(b^2 - 14b + 33)$
    $$= 10a(b - 11)(b - 3)$$

23. $w^2x^2 (y + 9)(y - 2)$

24. $2ax^2 - 2ax - 40a = 2a(x^2 - x - 20)$
    $$= 2a(x - 5)(x + 4)$$

25. $(2a + 3b)(2a - 3b)$

26. $3a^2b - 3a^2b^5 = 3a^2b(1 - b^4)$
    $$= 3a^2b(1 + b^2)(1 - b^2)$$
    $$= 3a^2b(1 + b^2)(1 + b)(1 - b)$$

27. $4rs^2(r - 3)(r - 9)$

28. $2y^2z + 38yz + 96z = 2z(y^2 + 19y + 48)$
$$= 2z(y + 3)(y + 16)$$

29. $(a^2 + b^2)(a + b)(a - b)$

30. $a^2x^4 - 81a^2 = a^2(x^4 - 81)$
$$= a^2(x^2 + 9)(x^2 - 9)$$
$$= a^2(x^2 + 9)(x + 3)(x - 3)$$

UNIT 20

1. $(7x + 3)(x + 1)$

2. $2y^2 + 5y - 3 = 2y^2 + 6y - y - 3$
$$= 2y(y + 3) - (y + 3)$$
$$= (y + 3)(2y - 1)$$

> sum $= 5$
> product $= -6$
> $+6$ and -1

3. $(7x - 3)(x - 1)$

4. $3x^3 - 5x^2 - 9x + 15 = x^2(3x - 5) - 3(3x - 5)$
$$= (3x - 5)(x^2 - 3)$$

5. $(2x + 7)(2x + 7)$

6. $6x^2 + 13x + 6 = 6x^2 + 9x + 4x + 6$
$$= 3x(2x + 3) + 2(2x + 3)$$
$$= (2x + 3)(3x + 2)$$

> sum $= 13$
> product $= 36$
> 9 and 4

7. $2(3 - a)(4 + a)$

8. $4x^3 - 10x^2 - 6x + 9 = 2x^2(2x - 5) - 3(2x - 3)$

The technique has shown that the expression cannot be factored. The expression is prime.

9. $(5x + 1)(x - 1)$

10. $4x^2 + 8x + 4 = 4(x^2 + 2x + 1)$
$$= 4(x + 1)(x + 1)$$

11. $(7x - 1)(x + 2)$

12. $2x^2 - 7x + 6 = 2x^2 - 4x - 3x + 6$

sum $= -7$
product $= 12$
-3 and -4

$\quad\quad\quad\quad = 2x(x - 2) - 3(x - 2)$
$\quad\quad\quad\quad = (x - 2)(2x - 3)$

13. $(2y - 7)(y - 5)$

14. $7x^2 + 32x - 15 = 7x^2 + 35x - 3x - 15$

sum $= 32$
product $= -105$
$+35$ and -3

$\quad\quad\quad\quad = 7x(x + 5) - 3(x + 5)$
$\quad\quad\quad\quad = (x + 5)(7x - 3)$

15. $2x(3w - 1)(w + 3)$

16. $6z^2 + 2z - 4 = 2(3z^2 + z - 2)$

sum $= 1$
product $= -6$
$+3$ and -2

$\quad\quad\quad\quad = 2[3z^2 + 3z - 2z - 2]$
$\quad\quad\quad\quad = 2[3z(z + 1) - 2(z + 1)]$
$\quad\quad\quad\quad = 2[(z + 1)(3z - 2)]$
$\quad\quad\quad\quad = 2(z + 1)(3z - 2)$

17. $z(xy - 8)^2$

18. $15x^2 - 2xy - 24y^2 = 15x^2 - 20xy + 18xy - 24y^2$

sum $= -2$
product $= -360$
-20 and $+18$

$\quad\quad\quad\quad = 5x(3x - 4y) + 6y(3x - 4y)$
$\quad\quad\quad\quad = (3x - 4y)(5x + 6y)$

19. $(3x + 2)(2x - 3)$

20. $2x^2 + 5x - 2$ $\quad$ prime

sum $= 5$
product $= -4$

21. $(4x - 3)(2x + 9)$

22. $xy^3 + 2y^2 - xy - 2 = y^2(xy + 2) - (xy + 2)$
$\quad\quad\quad\quad\quad\quad = (xy + 2)(y^2 - 1)$
$\quad\quad\quad\quad\quad\quad = (xy + 2)(y + 1)(y - 1)$

23. $(6x - 5)(2x + 1)$

24. $x^4 - y^4 = (x^2 + y^2)(x^2 - y^2)$
    $$= (x^2 + y^2)(x + y)(x - y)$$

25. $(1 + a^2)(1 + a)(1 - a)$

UNIT 21

1. $x = -7$ or $x = 2$

2. $x^2 + 13x + 30 = 0$
   $(x + 10)(x + 3) = 0$
   $x + 10 = 0$ or $x + 3 = 0$
   $x = -10$ $x = -3$

3. $x^2 - x + 7 = 0$
   Prime. There is no real solution.

4. $4x^2 + 8x + 4 = 0$
   $4(x^2 + 2x + 1) = 0$
   $4(x + 1)(x + 1) = 0$
   $x + 1 = 0$
   $x = -1$

5. $x = 0$ or $x = -5$

6. $x^2 + 2x = 8$
   $x^2 + 2x - 8 = 0$
   $(x + 4)(x - 2) = 0$
   $x + 4 = 0$ or $x - 2 = 0$
   $x = -4$ $x = 2$

7. $t = -5$

8. $5x^2 - 5x = 0$
   $5x(x - 1) = 0$
   $5x = 0$ or $x - 1 = 0$
   $x = 0$ $x = 1$

9. $x = \dfrac{1}{2}$ or $x = 3$

10. $2x^2 + 8x + 6 = 0$

    $2(x^2 + 4x + 3) = 0$

    $2(x + 1)(x + 3) = 0$

    $x + 1 = 0$ or $x + 3 = 0$

        $x = -1$ $x = -3$

11. $z = 3$ or $z = -7$

12. $10x - 10 = 19x - x^2$

    $x^2 - 9x - 10 = 0$

  $(x - 10)(x + 1) = 0$

    $x - 10 = 0$ or $x + 1 = 0$

       $x = 10$ $x = -1$

13. $x = 0$ or $x = \dfrac{-2}{3}$

14. $2 - 2x^2 = 0$

        $2 = 2x^2$

        $1 = x^2$

     $\pm 1 = x$

  $x = 1$ or $x = -1$

15. $w = \dfrac{1}{2}$ or $w = -4$

16. $2x^2 + 7x - 15 = 0$

  $(2x - 3)(x + 5) = 0$

    $2x - 3 = 0$ or $x + 5 = 0$

       $2x = 3$ $x = -5$

         $x = \dfrac{3}{2}$

17. $a = \dfrac{-1}{2}$ or $a = \dfrac{3}{5}$

18. $2x^2 - 5x - 3 = 0$

  $(2x + 1)(x - 3) = 0$

    $2x + 1 = 0$ or $x - 3 = 0$

       $2x = -1$ $x = 3$

        $x = \dfrac{-1}{2}$

19. $x = \frac{-4}{3}$ or $x = \frac{4}{3}$

20. $12x^2 + 5x - 2 = 0$
       $(4x - 1)(3x + 2) = 0$
       $4x - 1 = 0$ or $3x + 2 = 0$
       $4x = 1$ $3x = -2$
       $x = \frac{1}{4}$ $x = \frac{-2}{3}$

UNIT 22

1. $x = \dfrac{-3 \pm \sqrt{13}}{2}$

2. $\bigcirc\!\!\!2x^2 \bigcirc\!\!\!-3x \bigcirc\!\!\!-2 = 0$
    $\bigcirc\!\!\!ax^2 \bigcirc\!\!\!+bx \bigcirc\!\!\!+c = 0$
    Thus $a = 2$, $b = -3$, $c = -2$.

    $x = \dfrac{-b \pm \sqrt{b^2 - 4ac}}{2a}$

    $= \dfrac{-(-3) \pm \sqrt{9 - 4(-4)}}{2(2)}$ since $ac = 2(-2) = -4$

    $= \dfrac{3 \pm \sqrt{9 + 16}}{4}$

    $= \dfrac{3 \pm \sqrt{25}}{4}$

    $= \dfrac{3 \pm 5}{4}$

    The two solutions are:

    $x = \dfrac{3 + 5}{4},$ $x = \dfrac{3 - 5}{4}$

    $= 2$ $= -\frac{1}{2}$

3. $x = \dfrac{-1 \pm \sqrt{5}}{2}$

4. $2x^2 - 3x - 1 = 0$

   Thus $a = 2$, $b = -3$, $c = -1$.

   $$x = \frac{-b \pm \sqrt{b^2 - 4ac}}{2a}$$

   $$= \frac{-(-3) \pm \sqrt{9 - 4(-2)}}{4} \quad \text{since } ac = 2(-1) = -2$$

   $$= \frac{3 \pm \sqrt{9 + 8}}{4}$$

   $$= \frac{3 \pm \sqrt{17}}{4}$$

5. $x = 4$ or $x = -5$

6. $5(\frac{1}{5}x^2 - 5x + 1 = 0)$

   $x^2 - 25x + 5 = 0$

   Thus $a = 1$, $b = -25$, $c = 5$.

   $$x = \frac{-(-25) \pm \sqrt{(-25)^2 - 4(5)}}{2}$$

   $$= \frac{25 \pm \sqrt{625 - 20}}{2}$$

   $$= \frac{25 \pm \sqrt{605}}{2}$$

7. $x = 4$ or $x = -1$

8. $x^2 - 2x + 2 = 0$

   Also, $a = 1$, $b = -2$, $c = 2$.

   $$x = \frac{-(-2) \pm \sqrt{4 - 4(2)}}{2}$$

   $$= \frac{2 \pm \sqrt{-4}}{2}$$

   There is no real solution. We cannot take the square root of a negative number in the reals.

9. $x = 0$ or $x = \frac{1}{2}$

10. $4x^2 - 4x + 1 = 0$

    Also, $a = 4$, $b = -4$, $c = 1$.

    $$x = \frac{-(-4) \pm \sqrt{16 - 4(4)}}{2(4)}$$

    $$= \frac{4 \pm \sqrt{0}}{8}$$

    $$= \frac{1}{2} \quad \text{since } \sqrt{0} = 0$$

11. There is no real solution.

12. $x^2 - 2x - 10 = 0$

    Also, $a = 1$, $b = -2$, $c = -10$.

    $$x = \frac{-(-2) \pm \sqrt{4 - 4(-10)}}{2}$$

    $$= \frac{2 \pm \sqrt{44}}{2}$$

    $$= \frac{2 \pm 2\sqrt{11}}{2} \quad \text{since } \sqrt{44} = \sqrt{4 \cdot 11} = 2\sqrt{11}$$

    $$= \frac{\cancel{2}(1 \pm \sqrt{11})}{\cancel{2}}$$

    $$= 1 \pm \sqrt{11}$$

13. $x = \dfrac{-1 \pm \sqrt{37}}{6}$

14. $2x^2 + \sqrt{3}x - 4 = 0$

    Also, $a = 2$, $b = \sqrt{3}$, $c = -4$.

    $$x = \frac{-\sqrt{3} \pm \sqrt{3 - 4(-8)}}{4} \quad \text{since } b^2 = (\sqrt{3})^2 = (3^{1/2})^2 = 3$$

    $$= \frac{-\sqrt{3} \pm \sqrt{35}}{4}$$

UNIT 23

1. $x = \frac{1}{3}$ $\qquad$ $x = -2$ $\qquad$ $x = 5$

2. $x^3 + 3x^2 - 10x = 0$
   $x(x^2 + 3x - 10) = 0$
   $x(x + 5)(x - 2) = 0$
   $x = 0$ $\qquad$ $x + 5 = 0$ $\qquad$ $x - 2 = 0$
   $x = 0$ $\qquad\qquad$ $x = -5$ $\qquad\qquad$ $x = 2$

3. $x = -1$ $\qquad$ $x = 7$ $\qquad$ $x = 3$

4. $\qquad\qquad$ $x^3 - 49x = 0$
   $\qquad\qquad$ $x(x^2 - 49) = 0$
   $\qquad$ $x(x + 7)(x - 7) = 0$
   $x = 0$ $\qquad$ $x + 7 = 0$ $\qquad$ $x - 7 = 0$
   $x = 0$ $\qquad\qquad$ $x = -7$ $\qquad\qquad$ $x = 7$

5. $x = 0$ $\qquad$ $x = -8$

6. $\qquad\qquad$ $10x^3 - 3x^2 - x = 0$
   $\qquad\qquad$ $x(10x^2 - 3x - 1) = 0$
   $\qquad\qquad$ $x(5x + 1)(2x - 1) = 0$
   $x = 0$ $\qquad$ $5x + 1 = 0$ $\qquad$ $2x - 1 = 0$
   $x = 0$ $\qquad\qquad$ $5x = -1$ $\qquad\qquad$ $2x = 1$
   $\qquad\qquad\qquad\qquad$ $x = -\frac{1}{5}$ $\qquad\qquad$ $x = \frac{1}{2}$

7. $x = 0$ $\qquad$ $x = -2$ $\qquad$ $x = -3$

8. $3ax^3 - 24ax^2 - 99ax = 0$
   $\qquad$ $3ax(x^2 - 8x - 33) = 0$
   $\qquad$ $3ax(x - 11)(x + 3) = 0$
   $3ax = 0$ $\qquad$ $x - 11 = 0$ $\qquad$ $x + 3 = 0$
   $\qquad$ $x = 0$ $\qquad\qquad$ $x = 11$ $\qquad\qquad$ $x = -3$

9. $x = 5$ $\qquad$ $x = -4$

10.
$$2x^3 - x^2 + 14x - 7 = 0$$
$$x^2(2x - 1) + 7(2x - 1) = 0$$
$$(2x - 1)(x^2 + 7) = 0$$

$2x - 1 = 0 \qquad x^2 + 7 = 0$

$2x = 1 \qquad$ no solution to this equation

$x = \frac{1}{2} \quad$ is the only answer.

11. $x = -1 \qquad x = 1 \qquad x = -2$

12.
$$4x^3 + 18x^2 + 14x = 0$$
$$2x(2x^2 + 9x + 7) = 0$$
$$2x(2x + 7)(x + 1) = 0$$

$2x = 0 \qquad 2x + 7 = 0 \qquad x + 1 = 0$

$x = 0 \qquad\quad 2x = -7 \qquad\quad x = -1$

$$x = -\frac{7}{2}$$

13. $x = -4$

14.
$$x^6 - 64 = 0$$
$$(x^3 + 8)(x^3 - 8) = 0$$

$x^3 + 8 = 0 \qquad x^3 - 8 = 0$

$x^3 = -8 \qquad\quad x^3 = 8$

$x = -2 \qquad\quad x = 2$

15. $x = -2 \qquad x = 1$

16.
$$x^2 + cx - 3x - 3c = 0$$
$$x(x + c) - 3(x + c) = 0$$
$$(x + c)(x - 3) = 0$$

$x + c = 0 \qquad x - 3 = 0$

$x = -c \qquad\quad x = 3$

17. Not easily factored—at most, three real solutions.

18.
$$x^4 - 81 = 0$$
$$(x^2 + 9)(x^2 - 9) = 0$$
$$(x^2 + 9)(x + 3)(x - 3) = 0$$

$x^2 + 9 = 0 \qquad x + 3 = 0 \qquad x - 3 = 0$

$x^2 = -9 \qquad\quad x = -3 \qquad\quad x = 3$

no solution
to this part

The solution to this fourth-degree equation is $x = 3$ or $x = -3$.

19. $x = 3$ $x = -3$

20. $3x^3 + 2x^2 - 12x - 8 = 0$

$x^2(3x + 2) - 4(3x + 2) = 0$

$(3x + 2)(x^2 - 4) = 0$

$(3x + 2)(x + 2)(x - 2) = 0$

$3x + 2 = 0$ $x + 2 = 0$ $x - 2 = 0$

$3x = -2$ $x = -2$ $x = 2$

$x = -\frac{2}{3}$

UNIT 24

1.

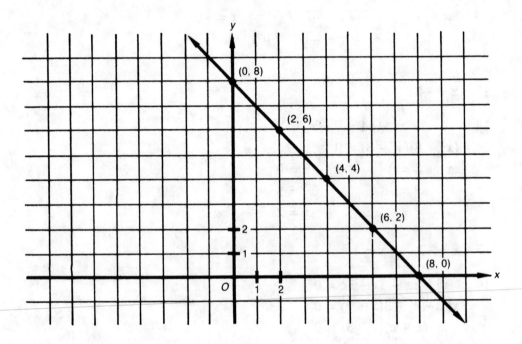

The x-intercept is 8.
The y-intercept is 8.

2.

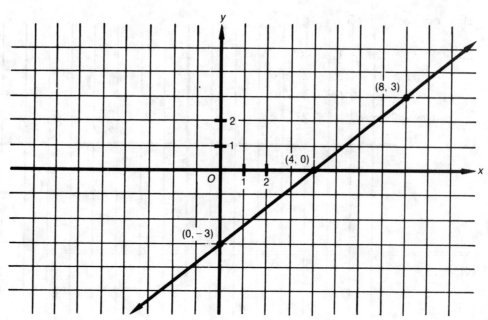

$3x - 4y = 12$

The x-intercept is where $y = 0$: $3x - 4(0) = 12$

$$3x = 12$$

$$x = 4$$

Therefore the x-intercept is 4.

The y-intercept is where $x = 0$: $3(0) - 4y = 12$

$$-4y = 12$$

$$y = -3$$

Therefore the y-intercept is –3.

3.

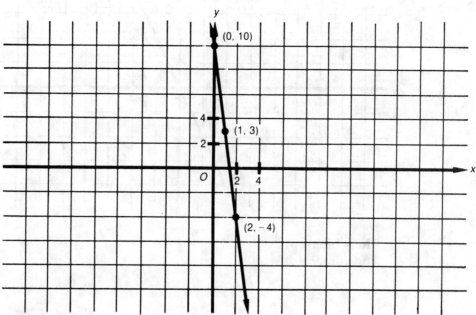

The x-intercept is $\dfrac{10}{7}$.

The y-intercept is 10.

4.

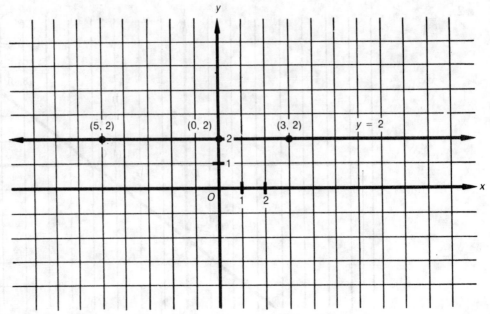

There is no *x*-intercept.
The *y*-intercept is 2.

5.

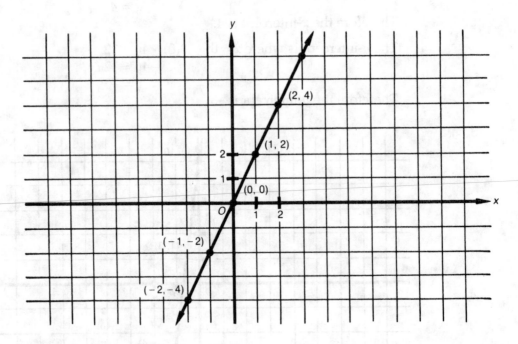

The *x*-intercept is 0.
The *y*-intercept is 0.

6.

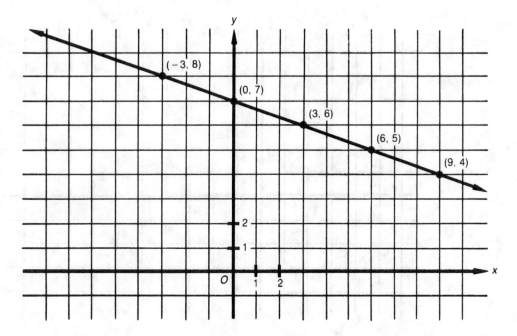

$$x + 3y = 21$$

$$x + 3(0) = 21$$
$$x = 21; \quad \text{the } x\text{-intercept is 21.}$$

$$0 + 3y = 21$$
$$y = 7; \quad \text{the } y\text{-intercept is 7.}$$

7.

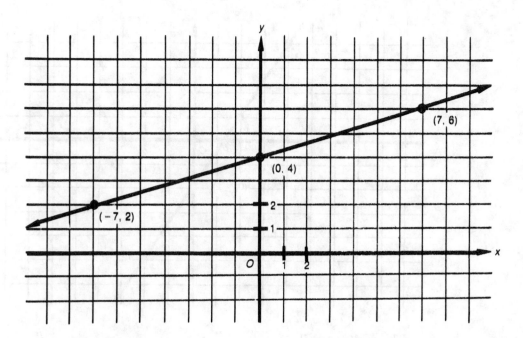

The x-intercept is –14.
The y-intercept is 4.

8.

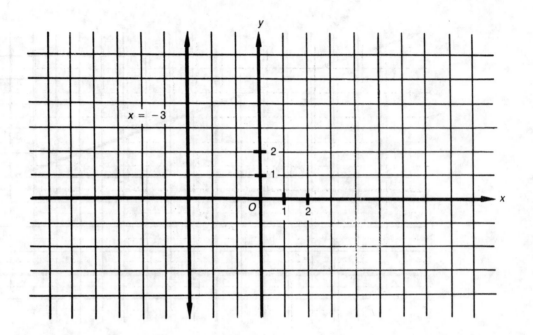

The *x*-intercept is –3.
There is no *y*-intercept.

9.

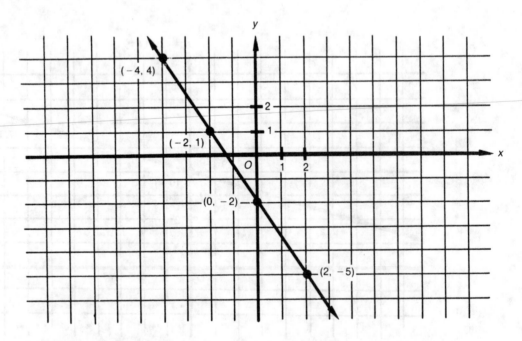

The *x*-intercept is $\frac{-4}{3}$.

The *y*-intercept is –2.

10.

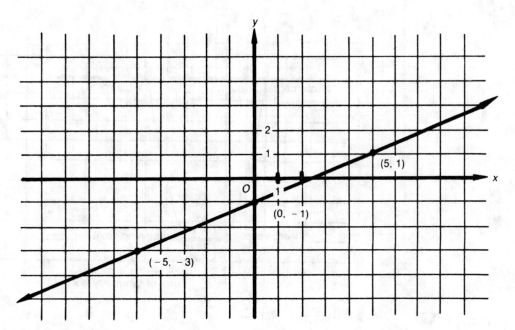

$$y = \frac{2}{5}x - 1$$

$$0 = \frac{2}{5}x - 1$$

$$0 = 2x - 5$$

$$-2x = -5$$

$$x = \frac{5}{2}$$

The x-intercept is $\frac{5}{2}$ or 2.5.

$$y = \frac{2}{5}(0) - 1$$

$$= -1$$

The y-intercept is –1.

UNIT 25

1.

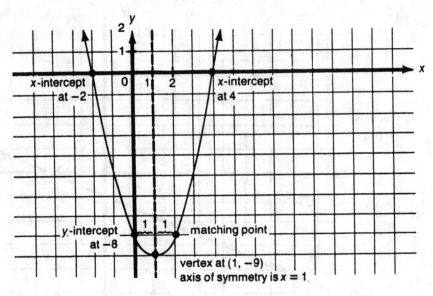

2. $y = x^2 + 2x - 8$ with $a = 1$, $b = 2$, $c = -8$

Parabola opens up.

y-intercept is -8.

x-intercepts: $\quad 0 = x^2 + 2x - 8$

$\qquad = (x + 4)(x - 2)$

$\qquad x = -4 \qquad \text{or} \qquad x = 2$

x-intercepts are -4 and 2.

Vertex: $\qquad \dfrac{-b}{2a} = \dfrac{-2}{2} = -1$

$\qquad$ If $x = -1$,

$\qquad y = (-1)^2 + 2(-1) - 8$

$\qquad = 1 - 2 - 8$

$\qquad = -9.$

$\qquad$ Vertex at $(-1, -9)$.

3.

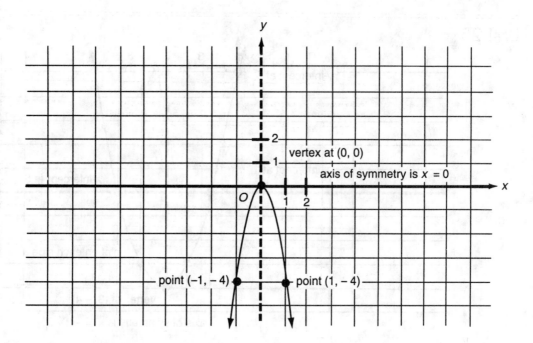

vertex at (0, 0)

axis of symmetry is $x = 0$

point (−1, − 4)

point (1, − 4)

4. $y = -x^2 - 4x - 3$ with $a = -1$, $b = -4$, $c = -3$
Parabola opens down.
y-intercept is –3.
x-intercepts:
$$0 = -x^2 - 4x - 3$$
$$= x^2 + 4x + 3$$
$$= (x + 3)(x + 1)$$
$$x = -3 \quad \text{or} \quad x = -1$$
x-intercepts are –3 and –1.

Vertex: $x = -2$, midway between x-intercepts
$$y = -(-2)^2 - 4(-2) - 3$$
$$= -4 + 8 - 3$$
$$= 1$$
Vertex at (–2, 1).

No need to find point on either side of vertex because we have x-intercepts.

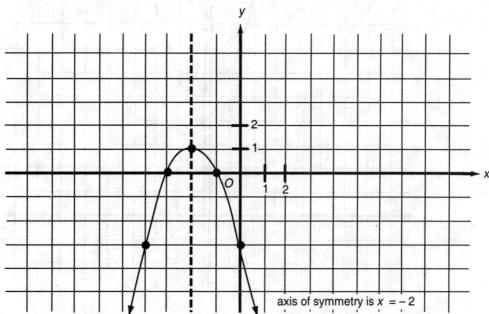

axis of symmetry is $x = -2$

5.

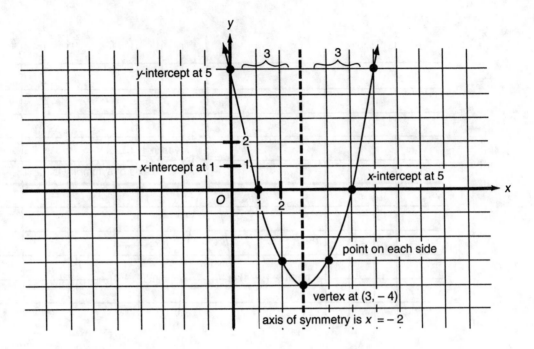

6. $y = 4 + 2x^2$ with $a = 2$, $b = 0$, $c = 4$

Parabola opens up.

y-intercept is 4.

x-intercepts: $\quad 0 = 4 + 2x^2$

$\qquad\qquad\qquad -4 = 2x^2$

No solution; there are no x-intercepts.

Vertex: $\qquad \dfrac{-b}{2a} = \dfrac{-0}{2(2)} = 0$

$\qquad\qquad y = 4 + 2\,(0)^2 = 4$

Vertex at $(0, 4)$.

Find one point on either side:

let $x = 1$; then $y = 4 + 2(1)^2 = 4 + 2 = 6 \quad (1, 6)$

let $x = -1$; then $y = 6 \qquad\qquad\qquad\qquad (-1, 6)$

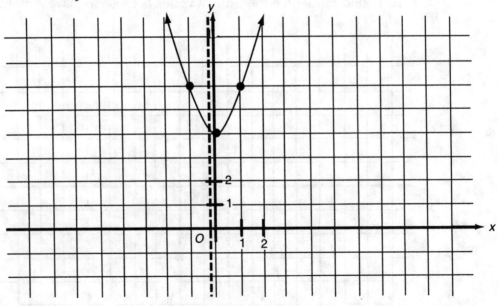

7.

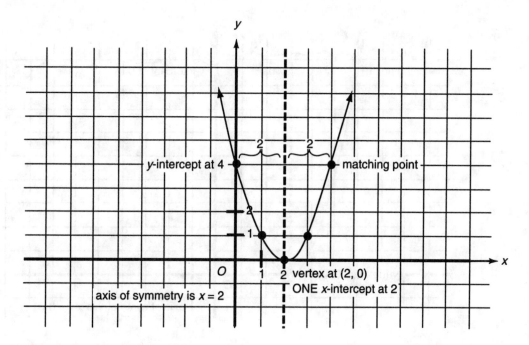

y-intercept at 4

matching point

axis of symmetry is $x = 2$

vertex at (2, 0)

ONE x-intercept at 2

8. $y = 25 - x^2$ with $a = -1$, $b = 0$, $c = 25$
Parabola opens down.
y-intercept is 25.
x-intercepts: $0 = 25 - x^2$
$\qquad\qquad\qquad = (5 - x)(5 + x)$
$\qquad\qquad$ x-intercepts are 5 and –5.

Vertex: $x = 0$, midway between x-intercepts
$\qquad\qquad$ $y = 25 - 0^2 = 25$
$\qquad\qquad$ Vertex at (0, 25).

Find one point on either side:
let $x = 1$; then $y = 25 - 1^2 = 24$ (1, 24)
let $x = -1$; then $y = 24$ (–1, 24)

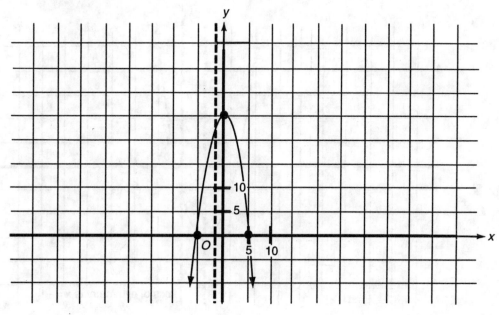

9.

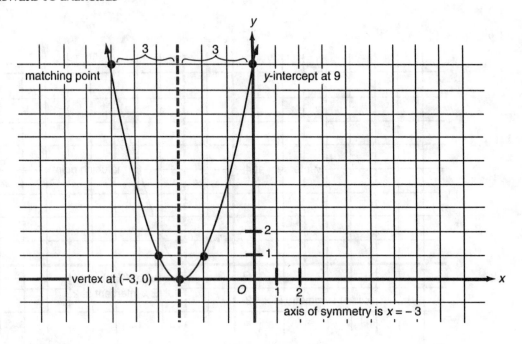

10. $y = 2x^2 + 4x - 1$ with $a = 2$, $b = 4$, $c = -1$

Parabola opens up.

y-intercept is -1.

x-intercepts: $0 = 2x^2 + 4x - 1$

Since the right-hand side is not easily factored, we will go to the next step.

Vertex: $\dfrac{-b}{2a} = \dfrac{-4}{2(2)} = -1$

$$y = 2(-1)^2 + 4(-1) - 1$$
$$= 2 - 4 - 1$$
$$= -3$$

Vertex at $(-1, -3)$.

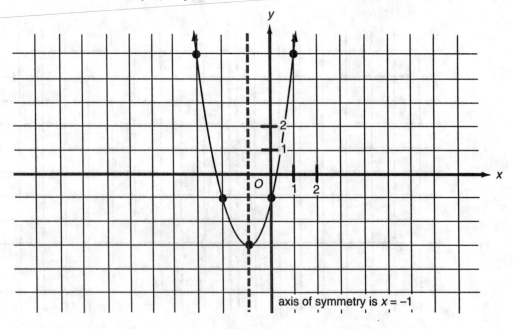

11.

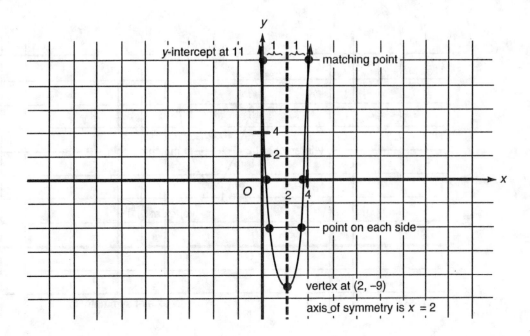

y-intercept at 11

matching point

point on each side

vertex at (2, –9)

axis of symmetry is $x = 2$

12. $y = -x^2 + 10x$ with $a = -1$, $b = 10$, $c = 0$

Parabola opens down.

y-intercept is 0.

x-intercepts: $0 = -x^2 + 10x$

$= -x(x - 10)$

x-intercepts are 0 and 10.

Vertex: $x = 5$, midway between x-intercepts

$y = -25 + 10(5) = 25$

Vertex at (5, 25).

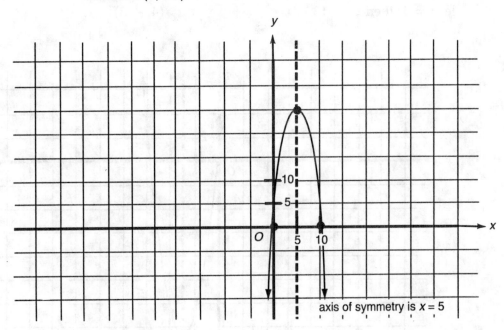

axis of symmetry is $x = 5$

13.

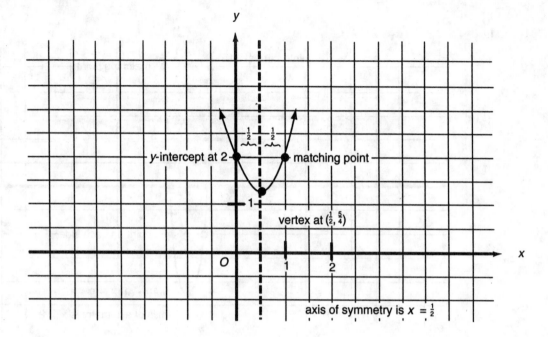

14. $y = 2x^2 - 12x + 3$ with $a = 2$, $b = -12$, $c = 3$

Parabola opens up.

y-intercept is 3.

Vertex: $\dfrac{-b}{2a} = \dfrac{-(-12)}{2(2)} = \dfrac{12}{4} = 3$

$y = 2(9) - 12(3) + 3 = -15$

Vertex at $(3, -15)$.

Find one point on either side:

let $x = 2$; then $y = 2(4) - 12(2) + 3 = -13$ $(2, -13)$

let $x = 4$; then $y = -13$ $(4, -13)$

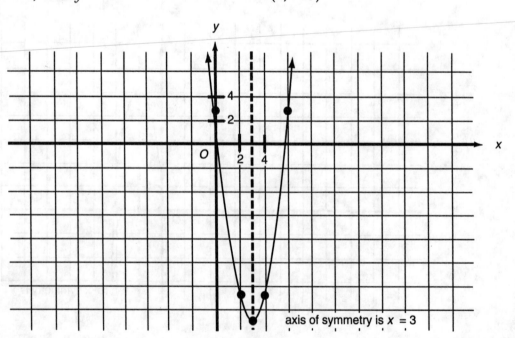

15.

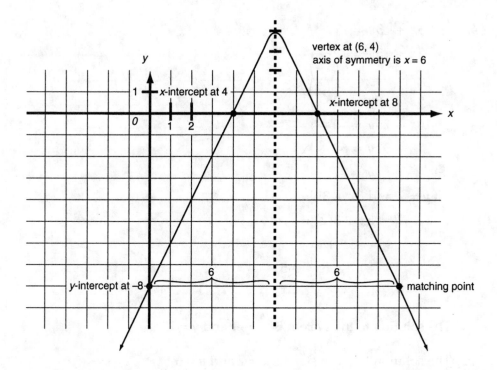

vertex at (6, 4)
axis of symmetry is x = 6

x-intercept at 4

x-intercept at 8

y-intercept at –8

6

6

matching point

UNIT 26

1. The solution to the system is $x = 4$ and $y = 1$.

2.
$$13x - 5y = -5$$
$$5(x + y = 1)$$

$$13x - 5y = -5$$
$$5x + 5y = 5$$
$$\overline{18x = 0}$$
$$x = 0$$

If $x = 0$ and $x + y = 1$,
$$0 + y = 1,$$
$$y = 1.$$

The solution to the system is $x = 0$ and $y = 1$.

3. The solution to the system is $x = -2$ and $y = -12$.

4. $+2(5x + 3y = 1$

   $5(-2x + 5y = 12)$

   $10x + 6y = 2$

   $\underline{-10x + 25y = 60}$

   $\quad\quad 31y = 62$

   $\quad\quad\quad y = 2$

   If $y = 2$ and $5x + 3y = 1$,

   $\quad\quad 5x + 3(2) = 1$,

   $\quad\quad\quad 5x + 6 = 1$,

   $\quad\quad\quad\quad 5x = -5$,

   $\quad\quad\quad\quad x = -1$.

   The solution to the system is $x = -1$ and $y = 2$.

5.   The solution to the system is $x = 2$ and $y = 6$.

6. $2x + y = 0$

   $\underline{\quad x - y = 1}$

   $3x \quad\quad = 1$

   $\quad\quad x = \frac{1}{3}$

   If $x = \frac{1}{3}$ and $x - y = 1$,

   $\quad\quad \frac{1}{3} - y = 1$,

   $\quad\quad\quad -y = \frac{2}{3}$,

   $\quad\quad\quad y = -\frac{2}{3}$.

   The solution to the system is $x = \frac{1}{3}$ and $y = -\frac{2}{3}$.

7.   The solution to the system is $x = 0$ and $y = 2$.

8. $11(-2x + 17y = 6)$

   $2(11x - 5y = -33)$

   $-22x + 187y = 66$

   $\underline{\quad 22x - 10y = -66}$

   $\quad\quad\quad 177y = 0$

   $\quad\quad\quad\quad y = 0$

If $y = 0$ and $-2x + 17y = 6$,

$$-2x + 0 = 6,$$

$$x = -3.$$

The solution to the system is $x = -3$ and $y = 0$.

9. The solution to the system is $x = 2$ and $y = 20$.

10. $4x + y = 13$
     $-4(x - 3y = 0)$

     $4x + y = 13$
     $-4x + 12y = 0$
     _____
     $13y = 13$
     $y = 1$

If $y = 1$ and $x - 3y = 0$,

$$x - 3(1) = 0,$$

$$x = 3.$$

The solution to the system is $x = 3$ and $y = 1$.

11. The solution to the system is $x = 2$ and $y = 7$.

12. $x + 4y \qquad = 8$
    $y = -\frac{1}{4}x - 7$

$x + 4y \qquad = 8$
$x + 4(-\frac{1}{4}x - 7) = 8 \quad$ by substitution
$x - x \qquad - 28 = 8$
$\qquad \qquad - 28 = 8$

There is no solution to the system.

13. The solution to the system is $x = 5$ and $y = 1$.

14. $2(2x + \frac{1}{2}y = 2)$

$\qquad 6x - y = 1$

$\qquad 4x + y = 4$

$\qquad \underline{6x - y = 1}$

$\qquad 10x \qquad = 5$

$\qquad x = \frac{1}{2}$

If $x = \frac{1}{2}$ and $6x - y = 1$,

$\qquad 6\left(\frac{1}{2}\right) - y = 1,$

$\qquad 3 - y = 1,$

$\qquad 2 = y.$

The solution to the system is $x = \frac{1}{2}$ and $y = 2$.

15. The solution to the system is $x = 1$ and $y = 12$.

16. $5(y = -\frac{2}{5}x + 4)$

$\qquad 5y = -2x + 20$

$\qquad 2x + 5y = 20$

$\qquad 3(x = \frac{1}{3}y - 7)$

$\qquad 3x = y - 21$

$\qquad 3x - y = -21$

$\qquad 2x + 5y = 20$

$\qquad 5(3x - y = -21)$

$\qquad 2x + 5y = 20$

$\qquad \underline{15x - 5y = -105}$

$\qquad 17x \qquad = -85$

$\qquad x = -5$

If $x = -5$ and $y = -\frac{2}{5}x + 4$,

$\qquad y = -\frac{2}{5}(-5) + 4,$

$\qquad = 2 + 4,$

$\qquad = 6.$

The solution to the system is $x = -5$ and $y = 6$.

17. There is no solution to the system.

18.
$$11x - y = 10$$
$$\underline{\quad 2x + y = \ 7}$$
$$13x \qquad = 17$$
$$x \qquad = \tfrac{17}{13}$$

If $x = \tfrac{17}{13}$ and $2x + y = 7$,

$$2(\tfrac{17}{13}) + y = 7,$$
$$\tfrac{34}{13} + y = 7.$$

Clear of fractions: $34 + 13y = 91,$
$$13y = 57$$
$$y = \tfrac{57}{13}$$

The solution to the system is $x = \tfrac{17}{13}$ and $y = \tfrac{57}{13}$.

19.   The solution to the system is $a = 2$ and $b = 2$.

20. $2x = -7(y + 1)$

$2x = -7y - 7$

$2x + 7y = -7$

Use substitution, since $y = -\tfrac{2}{7}x - 1$:

$$2x + 7y \qquad\qquad = -7$$
$$2x + 7(-\tfrac{2}{7}x - 1) = -7$$
$$2x - 2x \qquad -7 = -7$$
$$-7 = -7$$

There are infinitely many solutions to the system.

UNIT 27

1.   Use $x = 2$ to substitute:

$y = 3x^2 - 7x + 11$

$\quad = 3(4) - 7(2) + 11$

$\quad = 12 - 14 + 11$

$\quad = 9$

The solution to the system is $x = 2$ and $y = 9$ or $(2, 9)$.

2. Use $y = 4x^2$ to substitute:

$$-4x + y = 0$$
$$\downarrow$$

$$-4x + 4x^2 = 0$$
$$4x^2 - 4x = 0$$
$$-4x(x - 1) = 0$$

$$4x = 0 \qquad\qquad x - 1 = 0$$
$$x = 0 \qquad\qquad x = 1$$

If $\quad x = 0 \qquad$ If $\quad x = 1$

and $\quad y = 4x^2, \quad$ and $\quad y = 4x^2,$

$$y = 4(0) \qquad\qquad y = 4(1)$$
$$= 0. \qquad\qquad\quad = 4.$$

There are two solutions to the system.

$x = 0$ and $y = 0$ $\qquad\qquad$ (0, 0)

and

$x = 1$ and $y = 4$ $\qquad\qquad$ (1, 4)

3. There are two solutions to the system:

$x = 4$ and $y = 36$ $\qquad\qquad$ (4, 36)

and

$x = -11$ and $y = 176$ $\qquad\qquad$ (−11, 176)

4. Use $y = -x^2 - 1$ to substitute:

$$-x \quad + 5y \qquad\qquad = -5$$

$$-x \quad + 5(-x^2 - 1) = -5$$
$$-x \quad - 5x^2 - 5 \quad = -5$$
$$-5x^2 - x \qquad\quad = 0$$
$$5x^2 + x \qquad\quad = 0$$
$$x(5x + 1) \qquad = 0$$

$$x = 0 \qquad 5x + 1 = 0$$
$$5x = -1$$
$$x = -\tfrac{1}{5}$$

If $\qquad x = 0 \qquad$ If $\quad x = -\tfrac{1}{5},$

$$y = -x^2 - 1 \qquad y = -x^2 - 1$$
$$= 0 - 1 \qquad\quad = -(-\tfrac{1}{5})^2 - 1$$
$$= -1. \qquad\qquad = -\tfrac{1}{25} - 1$$
$$= -\tfrac{26}{25}.$$

There are two solutions to the system:

$x = 0$ and $y = -1$ $\qquad\qquad$ $(0, -1)$

and

$x = -\frac{1}{5}$ and $y = -\frac{26}{25}$ $\qquad\qquad$ $(-\frac{1}{5}, -\frac{26}{25})$

5. There are two solutions to the system:

$x = 3$ and $y = 3$ $\qquad\qquad$ $(3, 3)$

and

$x = -7$ and $y = -7$ $\qquad\qquad$ $(-7, -7)$

6. Use $y = x^2 + 2x + 5$ to substitute:

$$-4x + 2y \qquad = -7$$

$$\downarrow$$

$$-4x + 2(x^2 + 2x + 5) = -7$$
$$-4x + 2x^2 + 4x + 10 = -7$$
$$2x^2 = -17$$

There is no solution to the system.

7. The solution to the system is $x = 3$ and $y = 6$ or $(3, 6)$.

8. Use $y = 1$ to substitute:

$$y = 3x^2 + 8x - 2$$

$$\downarrow$$

$$1 = 3x^2 + 8x - 2$$
$$0 = 3x^2 + 8x - 3$$
$$0 = (3x - 1)(x + 3)$$

$3x - 1 = 0 \qquad\qquad x + 3 = 0$

$\quad 3x = 1 \qquad\qquad\qquad x = -3$

$\quad\; x = \frac{1}{3}$

There are two solutions to the system.

$x = \frac{1}{3}$ and $y = 1$ $\qquad\qquad$ $(\frac{1}{3}, 1)$

and

$x = -3$ and $y = 1$ $\qquad\qquad$ $(-3, 1)$

9. There is no solution to the system.

10. Use $y = \frac{1}{2}x^2 + x + \frac{5}{2}$ to substitute:

$$2x - y + 4 = 0$$
$$2x - \left(\frac{1}{2}x^2 + x + \frac{5}{2}\right) + 4 = 0$$
$$2x - \frac{1}{2}x^2 - x - \frac{5}{2} + 4 = 0$$
$$4x - x^2 - 2x - 5 + 8 = 0$$
$$-x^2 + 2x + 3 = 0$$
$$x^2 - 2x - 3 = 0$$
$$(x - 3)(x + 1) = 0$$
$$x - 3 = 0 \qquad x + 1 = 0$$
$$x = 3 \qquad x = -1$$

If $x = 3$ and $2x - y + 4 = 0$,
$$(2)3 - y + 4 = 0,$$
$$6 - y + 4 = 0,$$
$$10 = y.$$

If $x = -1$ and $2x - y + 4 = 0$,
$$2(-1) - y + 4 = 0,$$
$$2 = y.$$

There are two solutions to the system:

$x = 3$ and $y = 10$ \qquad\qquad (3, 10)

and

$x = -1$ and $y = 2$ \qquad\qquad (-1, 2)

11. The solution to the system is $x = 5$ and $y = 38$ or (5, 38).

12. Use $y = x^2 + 5x + 1$ to substitute:

$$y = x^2 + 3x + 9$$
$$x^2 + 5x + 1 = x^2 + 3x + 9$$
$$2x = 8$$
$$x = 4$$

If $x = 4$ and $y = x^2 + 5x + 1$,
$$y = (4)^2 + 5(4) + 1,$$
$$= 16 + 20 + 1,$$
$$= 37.$$

The solution to the system is $x = 4$ and $y = 37$ or (4, 37).

UNIT 28

1. $x \le 1$

2. $4 - (12 - 3x) \le -5$

    $4 - 12 - 3x \le -5$

    $-8 + 3x \le -5$

    $3x \le -5 + 8$

    $3x \le 3$

    $x \le 1$

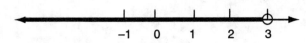

3. $x < 3$

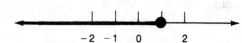

4. $4x + (3x - 7) > 2x - (28 - 2x)$

    $4x + 3x - 7 > 2x - 28 + 2x$

    $7x - 7 > 4x - 28$

    $3x > -21$

    $x > -7$

    $-7 < x$

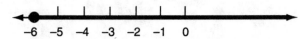

5. $x \ge -6$

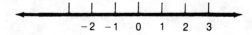

6. $3x + 4(x - 2) \ge x - 5 + 3(2x - 1)$

    $3x + 4x - 8 \ge x - 5 + 6x - 3$

    $7x - 8 \ge 7x - 8$

    $0 \ge 0$

The solution is the entire set of real numbers.

7. $x < \frac{-10}{3}$.

8. $3x - 2(x - 5) < 3(x - 1) - 2x - 11$

$3x - 2x + 10 < 3x - 3 - 2x - 11$

$x + 10 < x - 14$

$x - x < -14 - 10$

$0 < -24$

There is no solution.

9. The solution is the entire set of real numbers.

10. $5x - 2(3x - 4) > 4[2x - 3(1 - 3x)]$

$5x - 6x + 8 > 4[2x - 3 + 9x]$

$-x + 8 > 8x - 12 + 36x$

$-x + 8 > 44x - 12$

$-x - 44x > -12 - 8$

$-45x > -20$

$x < \frac{20}{45} \text{ or } \frac{4}{9}$

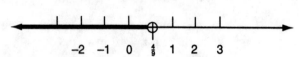

UNIT 29

The details for graphing each of the following parabolas can be found in the answers to the exercises for Unit 25.

1. $-7 < x < 3$

2. A rough sketch of $y = x^2 + 2x - 8$ is shown at the right. For what values of x is the curve **below** the x-axis?

    $-4 < x < 2$

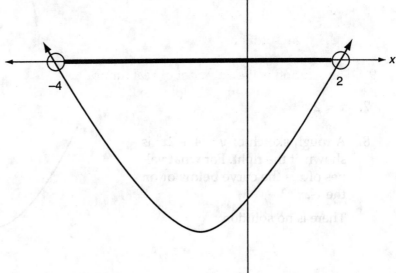

3. $x < -3$ or $4 < x$

4. A rough sketch of $y = -x^2 - 4x - 3$ is shown at the right. For what values of x is the curve **below** the x-axis?

    $x < -3$ or $-1 < x$

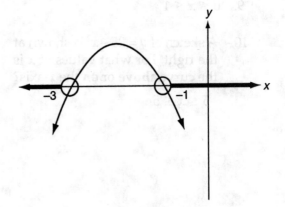

5. $-2 < x < 1$

6. A rough sketch of $y = -4x^2$ is shown at the right. For what values of x is the curve **above** or **on** the x-axis?

   $x = 0$

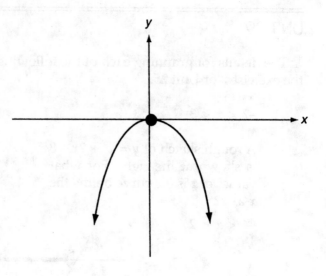

7. $x = 2$

8. A rough sketch of $y = 4 + 2x^2$ is shown at the right. For what values of x is the curve **below** or **on** the x-axis?

   There is no solution.

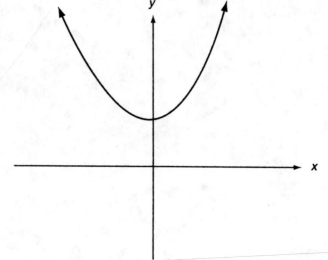

9. $0 \leq x \leq 1$

10. A sketch of $y = 25 - x^2$ is shown at the right. For what values of x is the curve **above** or **on** the x-axis?

    $-5 \leq x \leq 5$

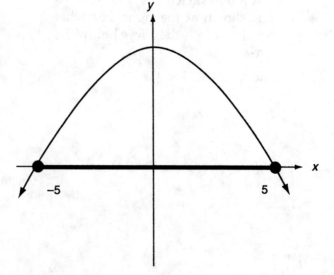

11. The solution is the set of all real numbers.

12. A sketch of $y = x^2 + 6x + 9$ is shown at the right. For what values of x is the curve **above** the x-axis?

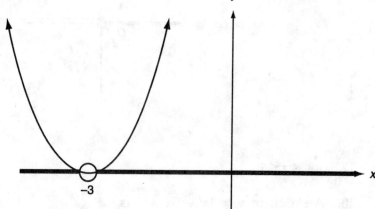

$x < -3$ or $-3 < x$ or,
in other words, all reals except -3

13. $x < -3$ or $5 \leq x$

14. A sketch of $y = 2x^2 + 4x - 1$ is shown at the right. For what values of x is the curve **below** the x-axis?

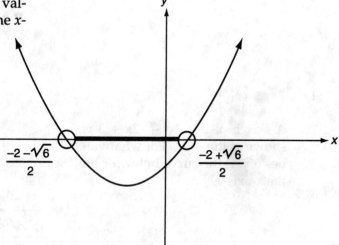

$$\frac{-2-\sqrt{6}}{2} < x < \frac{-2+\sqrt{6}}{2}$$

15. There is no solution.

16. A sketch of $y = -x^2 + 10x$ is shown at the right. For what values of x is the curve **below** or **on** the x-axis?

    $x \leq 0$ or $10 < x$

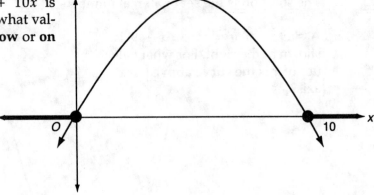

17. $-\frac{5}{2} \leq x \leq 1$

18. A sketch of $y = 3x^2 - 3x + 2$ is shown at the right. For what values of x is the curve **above** the x-axis?

    The solution is the set of all real numbers.

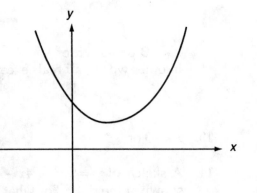

19. $-\frac{2}{3} \leq x \leq \frac{1}{2}$

20. A sketch of $y = 2x^2 - 12x + 3$ is shown at the right. For what values of x is the curve **below** or **on** the x-axis?

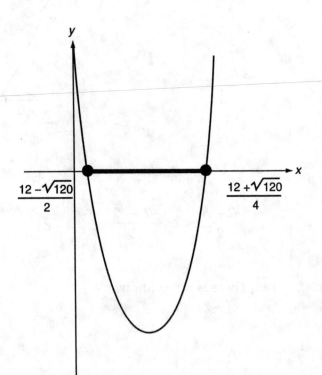

$$\frac{12 - \sqrt{120}}{4} \leq x \leq \frac{12 + \sqrt{120}}{4}$$

UNIT 30

1. $\log_3 81 = x$
    $3^x = 81$
    $= 3^4$
    $x = 4$

2. $\log_5 125 = x$
    $5^x = 125$
    $= 5^3$
    $x = 3$

3. $\log_7\left(\frac{1}{7}\right) = x$
    $7^x = \frac{1}{7}$
    $= 7^{-1}$
    $x = -1$

4. $\log 1 = x$
    $10^x = 1$
    $x = 0$

5. $\log_3 x = 2$
    $3^2 = x$
    $9 = x$

6. $\log_7 x = 0$
    $7^0 = x$
    $1 = x$

7. $\log_9 x = \frac{1}{2}$
    $9^{1/2} = x$
    $\sqrt{9} = x$
    $3 = x$

8. $\log_x 27 = 3$
    $x^3 = 27$
    $= 3^3$
    $x = 3$

9. $\log_x 49 = 2$
    $x^2 = 49$
    $= 7^2$
    $x = 7$

10. $\log_x 121 = 2$
    $x^2 = 121$
    $x = 11$

11. $\log 9 = \log 3^2$
    $= 2\log 3$
    $= 2(0.477)$
    $= 0.954$

12. $\log 8 = \log 2^3$
    $= 3\log 2$
    $= 3(0.301)$
    $= 0.903$

13. $\log 12 = \log(2^2 \cdot 3)$
    $= 2\log 2 + \log 3$
    $= 2(0.301) + (0.477)$
    $= 0.602 + 0.477$
    $= 1.079$

14. $\log\frac{2}{3} = \log 2 - \log 3$
    $= 0.301 - 0.477$
    $= -0.176$

15. $\log\frac{1}{2} = \log 1 - \log 2$
    $= 0 - 0.301$
    $= -0.301$

16. $\log\sqrt{3} = \log 3^{1/2}$
    $= \frac{1}{2}\log 3$
    $= \frac{1}{2}(0.477)$
    $= 0.2385$

17. $\log \sqrt[3]{2} = \log 2^{1/3}$

$\quad\quad = \frac{1}{3} \log 2$

$\quad\quad = \frac{1}{3}(0.301)$

$\quad\quad = 0.100$

18. $\log x^5 = 5 \log x$

19. $\log 2xy^3$

$\log 2 + \log x + \log y^3$

$\log 2 + \log x + 3 \log y$

20. $\log \dfrac{x^2}{y}$

$\log x^2 - \log y$

$2 \log x - \log y$

21. $\log \dfrac{x}{yz}$

$\log x - \log yz$

$\log x - (\log y + \log z)$

$\log x - \log y - \log z$

22. $\log \sqrt{x^3 y}$

$\log(x^3 y)^{1/2}$

$\frac{1}{2} \log x^3 y$

$\frac{1}{2}(\log x^3 + \log y)$

$\frac{1}{2}(3 \log x + \log y)$

$\frac{3}{2} \log x + \frac{1}{2} \log y$

UNIT 31

1.

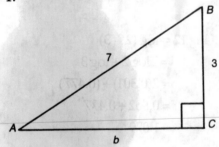

$c^2 = a^2 + b^2$

$7^2 = 3^2 + b^2$

$49 = 9 + b^2$

$40 = b^2$

$b^2 = 40$

$b = \sqrt{40}$

$\quad = 2\sqrt{10}$

2.

$c^2 = a^2 + b^2$

$\quad = 1^2 + 2^2$

$\quad = 1 + 4$

$\quad = 5$

$c = \sqrt{5}$

3.

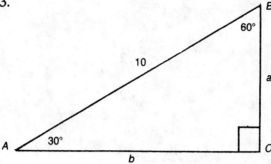

$$a = \frac{c}{2}$$
$$= \frac{10}{2}$$
$$= 5$$

$$b = \frac{c}{2}\sqrt{3}$$
$$= \frac{10}{2}\sqrt{3}$$
$$= 5\sqrt{3}$$

4.

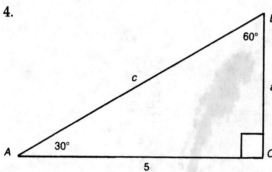

$$b = \frac{c}{2}\sqrt{3}$$
$$5 = \frac{c}{2}\sqrt{3}$$
$$10 = c\sqrt{3}$$
$$\frac{10}{\sqrt{3}} = c$$
$$a = \frac{c}{2}$$
$$= \frac{10}{\sqrt{3}} \cdot \frac{1}{2}$$
$$= \frac{5}{\sqrt{3}}$$

5.

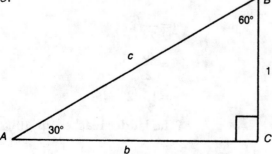

$$a = \frac{c}{2}$$
$$1 = \frac{c}{2}$$
$$2 = c$$
$$b = \frac{c}{2}\sqrt{3}$$
$$= \frac{2}{2}\sqrt{3}$$
$$= \sqrt{3}$$

6.

$a = b$

$6 = b$

$a = \dfrac{c}{2}\sqrt{2}$

$6 = \dfrac{c}{2}\sqrt{2}$

$12 = c\sqrt{2}$

$\dfrac{12}{\sqrt{2}} = c$

7.

$a = \dfrac{c}{2}\sqrt{2}$

$= \dfrac{10}{2}\sqrt{2}$

$= 5\sqrt{2}$

$b = 5\sqrt{2}$

8.

$a = \dfrac{c}{2}\sqrt{2}$

$= \dfrac{20}{2}\sqrt{2}$

$= 10\sqrt{2}$

The ladder reaches up $10\sqrt{2}$ ft.

9.

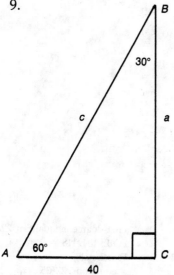

$c = 80$

$a = \dfrac{c}{2}\sqrt{3}$

$= \dfrac{80}{2}\sqrt{3}$

$= 40\sqrt{3}$

The building is $40\sqrt{3}$ ft. high.

10.

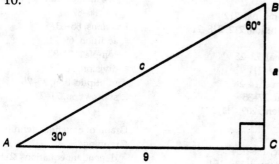

$b = \dfrac{c}{2}\sqrt{3}$

$9 = \dfrac{c}{2}\sqrt{3}$

$18 = c\sqrt{3}$

$\dfrac{18}{\sqrt{3}} = c$

$a = \dfrac{c}{2}$

$= \dfrac{18}{\sqrt{3}} \cdot \dfrac{1}{2}$

$= \dfrac{9}{\sqrt{3}}$

The observer is $9/\sqrt{3}$ ft. tall.

Index